The Plastic Turn

Ranjan Ghosh

Cornell University Press

Ithaca and London

First published 2022 by Cornell University Press

Printed in the United States of America

Library of Congress Cataloging-in-Publication Data

Names: Ghosh, Ranjan, author.
Title: The plastic turn / Ranjan Ghosh.
Description: Ithaca : Cornell University Press, 2022. |
    Includes bibliographical references and index.
Identifiers: LCCN 2022015191 (print) | LCCN 2022015192 (ebook) |
    ISBN 9781501766268 (hardcover) | ISBN 9781501766978 (paperback) |
    ISBN 9781501766282 (pdf) | ISBN 9781501766275 (epub)
Subjects: LCSH: Plastics—Philosophy. | Plasticity—Philosophy.
Classification: LCC B105.P537 G56 2022 (print) | LCC B105.P537 (ebook) |
    DDC 128—dc23/eng/20220602
LC record available at https://lccn.loc.gov/2022015191
LC ebook record available at https://lccn.loc.gov/2022015192

THE PLASTIC TURN

For Mahinder

CONTENTS

ILLUSTRATIONS

ACKNOWLEDGMENTS

The book has been a trial, and a fun and limit-stretching experience—a happy product in the end that owes a lot to a bunch of vibrant people who contributed in a variety of ways to its taking off, unfolding, and eventual polish and finish. It began as an idea four years ago and tested waters with insightful responses in the "mini-seminar series" that I was invited to deliver at the Critical Theory Institute, University of California, Irvine, in 2018. My host, Professor James Steintrager, the director of CTI, was splendid as was Dean Emeritus Georges Van Den Abbeele and Professor R. Radhakrishnan: Jim, Georges, and Radha, I cherish your friendship and the great conversations that we have had! The talk generated a lot of debate and instilled a fair proportion of intellectual struggle as "plastic turn" turned and twisted, wrote and rewrote itself for the next two years to become what it is now. In between, the trial continued as ideas and readings meshed and conflicted to produce "Plastic Literature" for the *University of Toronto Quarterly* 88.2 (2019): 277–291, "Plastic Controversy" in *Critical Inquiry's* "'In the Moment' portal" (2021), "Desiring-Material: Plastic-Art and Affect-ability," for *the minnesota review* 97 (2021): 53–76, and "The Plastic Turn" (2022) for the "Turn II" special number of *Diacritics*, 49.1 (2021): 67–87 (printed with permission from University of Toronto Press, Duke University Press, and Johns Hopkins University Press, respectively). A very special thank you to Andrea Bachner and Carlos Roja; Haun Saussy—always there with his support and thoughts. Thank you Bill Brown, Janell Watson, Ming Xie, and Jonathan Hart.

Wonderful and meaningful support came steadily from Daniel O' Hara (here and always for many things that we do together, my next door neighbor), Georges Van Den Abbeele (insightful and empathetic always), Claire Colebrook (brilliant and warm, exuding intellectual energy in everything we have done and have been doing), Catherine Malabou (generosity and erudition combined), Sidney Homan (blithe spirit, to say the least), Emily Apter (strong and supportive as ever), Caroline Rooney (always a colleague I can depend on), John Michael (sharp and soulful), Karen Pinkus (scholarship apart, how can I forget your assistance with research assistants?), Ethan Kleinberg (my true comrade), Heather Yeung (your book and our thoughts exchanged over Kafka, plasticity, and plastic literature), Marina Zurkow (Petroleum Manga connected us and continues to connect), and Rob Wilson (our friendship saw new wordlings always). This book brought me in close contact with outstanding artists with whom I investigated several areas of plastic material art. I owe you all a huge thank you: Tuula Närhinen, Judith Selby, Richard Lang, John Dahlsen, Evelyn Rydz, and Kelly Jazvac.

The brilliance and contribution of my research assistants—Madison Keele and Ecem Sarıçayır—changed the horizons of research that went into the book, canvassing readings and rare documents from the Cornell University Library and other sites. Beyond their research assistance, they stayed as my sustained support in formatting the book and preparing the index (Madison) and the bibliography in particular (Ecem), which, I must admit, just went out of hand through four years of intensive reading and research.

My editor at Cornell University Press, Mahinder Kingra (incidentally, the editorial director), was fantastic and relentless in his support for the project and the idea of the plastic turn. Insightful and critical, forthright and sharp, he steered the book throughout. Our many discussions over images, content, rewriting, and structuring of the project were remarkably productive, plasticizing the book to its final form.

The year 2020 has been a very distressful and difficult year with the COVID-19 pandemic changing our ways, habits, and values of living. Caught between two compelling forces, where the responsibilities as the head of the English Department, often, did not shake hands well with the joyous trial of finishing the book, I benefited immensely from the exemplary support of my colleagues: Pradipta Shyam Choudhury, Kaushani Mondal, Sumit Ray, and Binayak Roy. I cannot thank you enough for providing me the time and space to wrap the book up, and allowing me to focus on my writing when I needed it the most. Here I cannot miss mentioning Sukhendu Das for being so eager and generous with his assistance.

And, finally, I owe thanks to Sumana, always my first reader, for what most writers would not like to experience: hard criticism and parsimonious praise. I seriously start doubting the merit of my work when she signs off with an innocent and well collected "good." The manuscript went past her gate, this time, with a nod that I could not make much meaning of. This augurs well, I presume.

THE PLASTIC TURN

Turn to . . .

To elaborate (*travailler*) a concept is to vary both its extension and its intelligibility. It is to generalize it by incorporating its exceptions. It is to export it outside its original domain, to use it as a model or conversely to find it a model, in short it is to give to it, bit by bit, through ordered transformations, the function of a form.

—CATHERINE MALABOU

How can a *thing*—through its performance and operation, its poetics and politics of circulation—become an analogical and metaphorical axis for critical reflection and for thinking across disciplines and subjects not strictly related to the thing itself? It is on a note of relationality that an object may stimulate a theoretical formulation. Stefan Helmreich, for instance, borrows from Peter Galison, a historian of science, the concept of a theory machine to argue that mundane objects, under appropriate circumstances, can help generate complex theoretical structures. Among other examples, Helmreich notes that for physicist Sadi Carnot "water was a theory machine."[1] Zygmunt Bauman sees molecular and scalar properties of water in modernity. Peter Sloterdijk looks at globalization as foam. Hydrographic formations help the framing of ideas around a community, introduce issues of constructionism and mutability, and inspire dialectics between state, flux, and flow in identity formations. For

---

Epigraph from Malabou, *The Future of Hegel*, 7.

1. Helmreich, "Nature/Culture/Seawater," 132. See Galison, "Einstein's Clocks."

instance, "thinking with water" encourages relational thinking, as theories based on notions of fluidity, viscosity, and porosity reveal. Veronica Strang argues that the metaphoricity of water in hydrological thinking is inspired by "relations that are decidedly material." Water, she argues, "is a *matter* of relation and connection. Waters literally flow between and within bodies, across space and through time, in a planetary circulation system that challenges pretensions to discrete individuality. Watery places and bodies are connected to other places and bodies in relations of gift, transfer, theft, and debt. Such relationality inaugurates new life, and also the infinite possibility of new communities."[2]

Mielle Chandler and Astrida Neimanis see in water a mode of being called "gestationality"—a flow of thought that dismantles binary, "defies the either/or structure of activity and passivity, is neither active nor passive, and yet both active and passive." Here, water as theory machine is seen to have "a gestational orientation" that brings into existence that which is "not yet." Gestationality, thus, challenges "sovereign ontology," promoting "protoethical material phenomenon" and an "unpredictable plurality to flourish"—a kind of "aqueaous thinking."[3]

Surfing can also be seen in its metaphoricity and relationality. Deeply analogical to our critical thinking–ways across cultures, politics and traditions, wave riding—being intrinsically dependent on a host of forces (swell, wind, tide, sea-current, sand bar, weather system, pressure)—is an entangled and well-factored phenomenon but also "unreliable, inconsistent, and unstable." It argues out the notion of convergence and *agencement* representing a place that is in the state of becoming—webbed, meshed, and interactive.[4] Surfing, in its reconstitutive matrices, offers us a good critical-representational theorization of geopolitical culture and performatics of how we understand the place-space phenomena in contemporary discourses on globality.[5] So the surfed wave is interpreted as a relational space, as assemblage and convergence, and as a kind of "theorizing from the sea."[6] Helmreich notes that "scientific descriptions of water's form, molecular and molar, have become

2. Veronica Strang, "Introduction," 12 (emphasis in original).
3. See Chandler and Neimanis, "Water and Gestationality," 62.
4. "Agencement/Assemblage," 108–109.
5. Anderson, 'Relational Places," 576. See also Anderson, "Transient Convergence and Relational Sensibility."
6. Anderson, "Relational Places," 572.

prevalent in figuring social, political, and economic forces and dynamics."
He argues that "seawater has moved from an implicit to an explicit figure
for anthropological and social theorizing, especially in the age of globalization, which is so often described in terms of currents, flows, and circulations,"
and in the light of such tropes he suggests that "globalization might also be
called oceanization."[7] Proposing his "athwart theory," Helmreich points out
how theory can neither be set as above the "empirical nor as simply deriving
from it but, as crossing the empirical transversely."[8] Theory (and, for that
matter, seawater and, here, for me, plastic) is an abstraction, a materialization, and a *thing* in the world. Theories constantly cut across and complicate
our descriptive paths as we navigate forward in the real world. Be it water,
surfing, or plastic, we are in the midst of trying to decipher the syntax of materiality. So how athwart is plastic?

The analogical predication around a thing or an event (whether plastic
or surfing or water) builds with certain properties and attributions. Here,
if plastic is the word, plasticity is the "enclosure scheme": it is about seeing how
plastic concatenated the enclosure schemes of plasticity, plastic movements,
and plastic formations as found in critical-creative thinking in particular and
arts in general. It is the poesis of plastic-analogy. The plastic, in its materiality, both within and outside the laboratory, builds a somewhat equivocative
substitution, differentiations in expressions and distinctive conditions of understanding in relation to how the arts and thinking around man-material
interface develop and ramify. Analogical coevalness and co-occurrences are
the relevant points of concern and connect. They help us to distinguish between "the symmetrically analogous (by proportionality and denomination)
and the asymmetrically analogous (metaphors)."[9] It is the latter that deepens
the framework of my arguments.

Classically, analogy works through two classes—denominative analogy
(relational analogy) and analogy of proportionality (based upon the relations
of relations). The relational matrices within the context of my understanding work on the scale and axis that brings x (say, material) and y (say, the
aesthetic) not through a direct symmetry or equivalence, but, through intra-
schematism; x and y start to contrast each other, make sense and meaning

---

7. Helmreich, "Nature/Culture/Seawater," 133.
8. Helmreich, "Nature/Culture/Seawater," 134.
9. See Ross, *Portraying Analogy*, 86. See also Itkonen, *Analogy as Structure and Process*.

out of their asymmetricities into separate meaning-enclosures in the form of xy (the material overlapping the aesthetic, one substituting the other) and x-y (the material relationalizing the aesthetic without either of them losing their distinctive identities). Riding on equivocals (analogy is not about being analogous on all points of contrast because discriminative overlaps are always in order), relational analogy brings into the intrascheme contrast plastic as a material (x) and plastic as an aesthetic in itself (y). It is out of the asymmetry (plastic sea as against world literature or plastic polymerization as against modernist intertextuality, x-y) that meaning-relevance emerges, building on the conceptual and performative predication of each other.

Theory machine works on such meaning-relevance and asymmetrical metaphoric meaning-making. The material and aesthetic (plastic polymerization corresponding with artistic intertextuality), the matter and mattering (micro plastic-behavior analogized with literary materialization), and the object and the poetics of object-behavior (microplastic dissemination and its invasive ways gridded on to the complexities of world literature across nationalist and cultural borders), are implicative, associatively contrastive, and analytic. The material-aesthetic builds its own contexts for intrascheme enclosures of understanding. Explaining through denominative analogy, James F. Ross looks at the variation that representational denomination brings where x represents y and x functions like y and x is a picture of y. Plastic and thinking around the arts analogize on the coordinates of formation, processing, structure, behavior, and properties. There is a transposition of content—the material plastic with its content (x) asymmetrically contraposed with doing literature and arts (y) that have their distinct content-processing (x-y)—and indications, mostly hidden, generated through a certain line of perception, chain of association, and understanding.[10] The denominative (x as y), deductive (y drawn from x), representational (x like y), and (a)symmetrical (x-y and xy) are all part of the figural and metaphorical.

The turn for me hinges on the x-y and xy, where the hyphenation in particular brings about the analogical or metaphorical motor: plastic behavior (x) in its materiality, material formation, and mutation builds a productive hyphen with literary behavior and orientation (y), resulting in the material and the aesthetic conjoining through a vestibule of meaningful exchange as

---

10. See Zilberman, *Analogy in Indian and Western Philosophical Thought.*

represented through a hyphen (*x-y*): the material-aesthetic. This inhabitation in the hyphen opens up our thought-boxes—the supplementaries—beyond the mere simplistic post-Goethean understanding of plasticity as generative, transformative, and creative.

Plastic, in this book, is a discourse, at once conceptual and material, and it is an aesthetic figure that emerges from the material. As a material and material-problematic, plastic has its own structure of representations and meanings; however, as a discourse, it builds an oppositional network of concepts and signifiers. Adi Efal observes that

> plastically speaking, when one uses the term "figure," one usually refers to a situation in which exists a synchronous delineation of two surfaces, one containing, surrounding, enveloping or carrying the other, as in the following scheme: A figural situation denotes a gesture of framing, cutting out, a distinction of an outline of a surface or a platform, which implies necessarily the containment of one form by another form. Therefore, the figural dynamics contains a bilateral movement from a form to a platform and back from the platform to a form.[11]

The "material-figural" plastic, as part of a bilateral movement in which plastic keeps revising its status as a form and a platform, builds a plexus of oppositional thinking that routes and orients itself across its heterogenous existence within the laboratory and outside it. In a figural dynamics wherein the material contains the aesthetic and the aesthetic enframes the material, plastic demonstrates a coexistence within identities of different orders (chemo-eco-cultural) and within an incommensurability of emotions ranging from love and enthusiasm to shock and annoyance to helplessness and habituality. It is a *thought* and also continuity in *thinking*. Plastic, in this book, is a material form, discourse, representation, and negation of understanding, in both its materiality and its aesthetics of materialization. In its material presence, plastic becomes referential, designatory, and significatory. By radicalizing the material through the aesthetic and the aesthetic through the material and its characteral conditions, plastic exists as an *event* through the book: visible, haptic, diagrammatic or geometric, referential, structural, curvilinear, differential, and hence, figural.

---

11. Efal, "Gravity of a Figure," 39.

Plastics in their materiality—whether as microplastics in the sea or plastic mineralization through the landfill or plastic in a variety of forms washed up on the seashore—are loaned a language in forms that are semiotic and aesthetic. Working through Deleuze's *enoncable* or signaletic material, we find that the material-plastic can be made utterable through images and discourses that are affective, rhythmic, kinetic, and sensory.[12] Language conditions the non-language material (plastic) through the construction of an aesthetic; plastic, in the pages of this book, becomes "significatory" and articulative through metacommentaries on world literature, comparative literature, creative thinking, reflections on geo-formations, and various other discourses. It is here that the material-aesthetic (with the hyphen) emerges and takes a complicated life of its own: the material-signifier, the materiality of the signifier, and the material as signifier come together to form their discursive formations. In such a dialectic between the nonlanguage material and language as conditioning the expressive potential of the material, plastic becomes both the line and the letter, as well as a resonance, a material marker, and a code.

This dialectic also makes plastic a "material metaphor" where the axis of transference gets built between the material and sign-concept. The material plastic transforms itself to generate certain notions of materiality, which then produce their levels of conceptual-aesthetic signifying power.[13] Plastic figurality combines an interiority of its function and existence in forms that we can express through formulas and diagrams (as chapter 2 demonstrates) and an exteriority that complicates and diversifies representation and understanding of presence and operation (as enunciated in chapters 3, 4, and 5). The signification and meaning in both states of expression call for what I argue to be the "Plastic Turn," where the focus and emphasis is never outside the plastic, both in its voluble presence (in sight) and in its absence (out of sight). Plastic talks as a material in its material formations, in modes of structuration, and in metaphoric figurality that imports "excess signification" into our aesthetic understanding and critical thinking. My arguments in the book continually connect the first and the second order levels of meaning into a reciprocal space of negotiation where the material-plastic is argued to con-

---

12. See Deleuze, *Cinema 2*, 29.
13. See Boomen, "Interfacing by Iconic Metaphors." See also Hayles, "The Transformation of Narrative and the Materiality of Hypertext."

tribute to poetic plasticity—the material-metaphoricity. This is not smooth substitution of one with the other in a kind of master discourse. If plastic polymerization is interpreted as modernist and postmodernist intertextuality (chapter 2), one must not forget the differentiality of intertextuality as a strategy or phenomenon in literary-comparative studies. Hence, polymerization is commensurate with intertextuality, but also builds its points of discontinuity (the material-aesthetic as discriminative overlap; a discriminative relationality). It is on this point that plastic exceeds its materiality as a signifier (the material as code) to deliver or produce an excess, which I term "plasticity" (following the codes available for transgression and supplementation). Matter and mattering are not terms of simple equivalence and direct points of correspondence. Plasticity, therefore, for this book, is not a pre-materialistic phenomenon but a material-figural event that comes *after* the discovery of plastic; our consciousness is never without the plastic-material.

The material, in communication with an aesthetic, and the aesthetic, articulating a fresh understanding of the material, announce a synthetic movement that binds one concept to the "materiality of the matter" and allows an explication of the material-aesthetic. For instance, plastic polymerization can explain comparative intertextual literary practice (chapter 2) or a world and transcultural-transcontinental understanding of literature as it is figuralized in relation to the travel of microplastic in the sea within a compelling material-aesthetic that I call "Plastic Literature" (chapter 4). This synthetic movement is about plastic's implicative complexes with different frames and foci—plastic becomes an "emphatic metaphor."[14] The relationality and resonance of plastic metaphoricity generate "implicative system(s)." The matter—the elementality and mattering—has its optic and the semiotic, discursivity, and the haptic, product and the process, t(h)ere and not-(t)here, bringing out varied modes of utterance in literature, arts, cultures of thought, and eco-political circles of living and survival. Plastic, thus, "entangles different kinds of conceptual relations."[15] With different "motives for metaphor," we begin to think with and through plastic.

---

14. Black, "More about Metaphor," 442. Black notes that "it would be more illuminating in some of these cases [i.e., of metaphors imputing similarities, which are difficult to discern otherwise] to say that the metaphor creates the similarity than to say that it formulates some similarity antecedently existing" (451).

15. Schafer, "Historicizing Strong Metaphors," 35.

Figuring his "limits of fabrication," Nathan Brown argues that "both material science and materialist poetics engage the boundaries of formal invention and material construction in their respective fields. These fields are, to say the least, markedly different. But they are nevertheless related insofar as their boundary work draws them into a common terrain of formal, constructive, and ideological problems attendant upon experimental practices of making."[16] More about figuration and analogical-metaphorization than fabrication, this book reads into plastic behavior, its properties and structures, complexities of polymerization, chemo-poetics and aesthetics of chemistry, the impact and indulgence of plasticizers, microplastic dissemination and contamination and various other material issues to configure how we might start to read and experience poetry and paintings, anthropocenic and Earth art, world-comparative literature, the poetics of globalization and other critical concerns. It is plastico-poesis—not an undifferentiated poetics of plasticity—invested in the material, materialization, and mattering, and in their aesthetic potentializations and figurality. The plastic is both a phenomenon inside the laboratory and outside it—as a material, as and how it *matters*, its materialization, corresponding with how we think, write, reflect, and react. How can we in the "Plastic Turn" overturn issues from their conventional embeddings into vexed pathways of thinking when "plastic as theory" is about "looking on" and "contemplation" and a "seeing through"? Having its own object-features and behavioralities, plastic becomes wholesome enough to offer a culture of engagement, thinking, and discursivity. I qualify this as quantum theorization where an object, in its materiality, culturality, and metaphoricity, manifests conceptual and theoretical refigurations: a state of receptionist unfixation and an emergence that unhinges its own understanding. It is here that relationality matters—the poetics of transformative correspondences and ungroundedness. Relationality is configuring lines of *mattering* as they intersect, entangle, constellate, and trajectorize. The theory machine functions in "relational mattering": provisionality, permanence, plurality, and the precarity of plastic. Plastic as theory machine is inseparable from the diffractive, meaning-making potency of matter. It has "heterogenous history," presence, and it exists as a narrative: a weave, a nonliving *potenza*, an entanglement. Does plastic approximate (to use Karen Barad's term) "trans/materialities"—the literary-philosophical extensions and entanglements?[17] How do plastic-properties—the

---

16. Brown, *Limits of Fabrication*, 13.
17. Juelskjær and Schwennesen, "Intra-Active Entanglements," 16.

more-than-human liveliness and potency—become synonymous with literary-philosophical performativity? Plastic then is both theory and theorization.

This book is neither exclusively on plastic nor a philosophical elaboration on the principles of plasticity. As part of our petromodernity, plastic fixes and unfixes the energy regimes that expose us to certain conditions of materiality, sociocultural conditions of consumption, the politics of garbology, commodification, and desire, and a unique globalization of need and disposability. As extraction and extrusion, material acculturation and submission, egalitarian utility and waste-power, plastic (as oil-detritus) speaks in the language of profit and loss, a petrochemical living, survival, sacrifice, and extinction zone. Plastic, within certain eco-techno-cultural systems, builds its material points of expression, and, with them, generates new forms of value-making. So the denominative and inevitable narratives emerging out of the proposal for a plastic turn bring an obvious and expected retinue of issues that have been explored elsewhere in a variety of books and papers: the philosophy of plasticity, the cultural history of the material plastic and its impact, plastic pollution as it harms our physiological systems, transforming sexualities across species, and changing our clinical identities, the fall-outs of climate change and anthropocenic setbacks, disastrous marine pollution, and the plastic economy as it brings its own discourses on capitalism, precarity, and global capital flow. *The Plastic Turn* is not about these issues; nor is it about ecological activism, anti-plastic campaigns and forms of globalization as revealed through capital generation, exploitation, and peripheral zones of sustenance within a labyrinthine petro-culture. None of these is the turn to plastic for the book.

What kind of plastic does the book turn to? What is the Plastic Turn that it declares? I propose a plastic turn that does not emphasize the formal quality of plasticity but takes inspiration from the materiality of plastic itself. It is on the form of plastic and the material-aesthetic platform of plasticity and non-plasticity of plastic. The *turn* is plastic on two counts: first, plastic came into being as a material for ready and daily use, with an extraordinary reach, variety, and efficacy. And second, most importantly, I argue that the emergence of plastic as a material brought forth what I call the "material-philosophical-aesthetic"; for me, this means the chemical-organic-material-experimental-global-structural-technological outgrowth and ramification of plastic correspond, theoretically and organically, with our thinking in arts, metaphysics, ecologics, poetics, and literature throughout the twentieth century. The formation, happening, and performance of plastic correspond with the configuration, consequence, and concepts that permeated our thinking in literature and the arts. I see the growth (as a product in the culture industry) and formation (substance-variety through multiple application and chemical synthesis) and dissemination (oceanic movements and sea-land drifts and percolations) of plastic with corresponding developments in literature and critical thinking beginning with the turn of the twentieth century.

The anchors are dropped deep and wide on material plastic. *The Plastic Turn* never slips away from the material, and it keeps exploring its plasticity and nonplasticity as much as its material-affective-aesthetic power and purpose that both staying in touch with the material and being touched by the material bring. The book is on plastic both within the laboratory (its polymericity and molecularity) and outside it (its metamorphosis and motility) but not without its aesthetic projections and formations in either domain of existence. For instance, how the composition and chemicality of plastic material-aestheticizes the formation and formability of literary text and reading: the chemo-poetics of understanding (Turn 1); the turn to the material and its aestheticization (Turn 2); how the distribution, dissemination, and drift of marine plastic correspond with and figuralize the understanding of comparative and world poetics and literature: plastic literature (Turn 3); how the discovery of plastiglomerates (rocks with molten plastics in them) raises issues across a disturbing bandwidth involving what I call plastifossilization, geo-art, plastic sublime, geological imagination, and psycho-physiological disrupture—the material-aesthetic coordinates of our plastic existence (Turn 4).

The Plastic Turn is then a co-occurrence and collateral poesis of the material with the aesthetic of critical thinking across diverse disciplines in the twentieth century. The book breathes and settles on the hyphen that constructs the material-aesthetic, expansively bringing into prominence four major turns but not without the other *turns* that inform and inflect the book's dominant circulations: plastic grows before us visibly and builds an invincible and invisible presence that leaves us with a kind of attunement, a consciousness about plastic existing here, there, and then *somewhere*. When plastic arrives after we have made practical and convenient use of the material, we encounter a kind of strangeness, unbeknownst to the life-web we survive and are sustained in. This is the paradigm of the unpresentability of plastic: the aesthetic of the yet-to-come plastic, yet-to-form plastic, and transforming plastic. Are we confronted with a strange communication between plastic nature and plastic culture? Is plastic nature an outlawed nature? Has plastic "enfleshed" the world? What affective fallacies does it bring? How can we conceive plastic-art? How has plastic repremised our notions of finitude? Has the Plastic Turn declared an ontology of its own that is never without the possibilities—im/permanence—of its ontological re-formulation? Many such questions—not limited to the ones spelled out here—inform the book's material-aesthetic axis as the chapters keep adding turns to the four major turns that ground and girdle the book-core.

Plastic has built a new sensibility: the search for the new form and man-world-material connect introduces a fresh perspective of reality. This is a plastic-induced reality that brings a new understanding of situations, in their cultural, psycho-social, ecocritical viability and pragmatism and heuristic processes of expression and language. In its molecularity, polymericity, and molarity, plastic points to a turn in twentieth-century thinking, in thought-differentialities and critical understanding across disciplines and discourses. It is the *material-aesthetic* as the operative theory machine.

Chapter 1

The Plastic Turn

Plastics happen; that is all we need to know on earth.
—Richard Powers, *Gain*

Material-Aesthetic I

The careers of plastic-problematic and plastic-performative begin with a paper read before the Society of Arts by Mr. Alexander Parkes of Birmingham on December 20, 1865, explaining what Parkesine means:

> For more than twenty years the author entertained the idea that a new material might be introduced into the arts and manufactures, and in fact was much required; he succeeded in producing a substance partaking in a large degree of the properties of ivory, tortoiseshell, horn, hard wood, India rubber, gutta percha, etc., and which will, he believes, to a considerable extent, replace such materials, being capable of being worked with the same facility as metals and wood. This material was first introduced under the name of parkesine (so called after its inventor), in the Exhibition of 1862, in its rough state, and manufactured into a variety of articles in general use.[1]

---

1. As quoted in Gloag, "The Influence of Plastics on Design," 463; see Parkes's "On the Properties of Parkesine."

Parkesine was a composite—a "material event," as it were, that partook of a variety of things and underwent a complex process in hardness and flexibility working through "pyroxyline and oil, alone or in combination with other substances."[2] It was after the invention of Bakelite at the turn of the nineteenth century that the *turn* to plastic announced itself in the second decade of the twentieth century. Plastic, as it chemically and structurally evolved and proliferated, aroused great interest. It was given generous space in an important exhibition of British Art in Industry and promoted by the Royal Society of Arts in collaboration with the Royal Academy in 1935.[3] An exhibit labeled "Resin M" uncorked serious enthusiasm—the colorless object was found to be transparent, with a glass-like sparkle without being glass, and a dramatic presence that provoked the imagination and medley strains of thought. Plastics, both thermoplastic (cellulose nitrate, cellulose acetate, acrylic resins, polystyrene) and thermosetting (phenolic resins, cast phenolic, urea resins), unconcealed their seduction and invoked further interest through attributes of molding and a malleability that raised possibilities of an infinite number of shapes and forms and, hence, concepts. Plastic initiated a turn in thinking, habits, lifestyles, emotions, economy, and passion.

The scarcity of silk and ivory brought a turn to plastic within advanced capitalism in the postwar period (after WWI and dominantly after WWII) and it never failed to fuel fantasy and the imaginary. The passion brought the performative into play: a restlessness over the new achievement sought to discredit existing materials such as glass, aluminum, wood, and other material alternatives. This triggered an enormous predilection to replace and displace *all* through/with plastic. The plastic turn brought "a world of color and bright shining surfaces, where childish hands will find nothing to break, or sharp edges and corners to cut or graze, no crevices to harbor dirt or germs."[4] With its ubiquity, plastic inaugurated a dream and played the "underminer": its use 'undermines the idea that materials possess their own genuine image since the outer skin no longer represents the core and the surface is free to exist in its own right, truly synthetic and tailored to meet specific needs."[5] Jeffrey Meikle, quoting Waldemar Kaempffert, the science editor of the *New York Times*, points out how the plastic phenomenon and performativity

2. Gloag, "The Influence of Plastics," 462.
3. Gloag, "The Influence of Plastics," 464.
4. Walker, "Plastics," 67.
5. Walker, "Plastics," 67.

promised nth versions and nth possibilities resulting in an expansionist re-gime that territorialized both the economic and popular landscape, a sort of plastic imperialism. It became a plastic fold-in: mobility, quick change, and impermanence were exciting and exhilarating to a point that everything became a possibility of being fashioned out of an "appropriate synthetic": "a thermoset world was melting into thermoplasticity."[6]

Plastic, through relentless chemo-synthetic transformations, has come to construct a complicated mobility with a variety in color, smoothness, and form, through its appearance-reality dialectic, and by being performatively sensorial-experimental. Transformations and variety manifest in the ways in which they express themselves, circulate through the economy, inform the habits of use and disposability, accredit their presence against others, and find their value and valence. Plastic has its own modernist restivity and allure, temptations and tests, and a progress in contrariety: it "represents a shiny new world, one that removes people from the cycles of life and death, one that supersedes the troublesome, leaky, amorphous, and porous demands of our ancestors, our bodies, and the earth. Ridding ourselves of the demands of the earth seemed to promise a world of prosperity through scientific control."[7] It brings about the interesting dialectic between control and ex-periment, expressive frenzy and plan-order, the dominance of the mind and the response-effect of the body, intelligence and intentionality, with purpose and programs—an agony of fine excess.

As a statement and style, plastic tries to establish a "wit" that is hegemonic and totalizing in nature; the relevance of its limits is nearly lost in the in-toxicative submission that it brings. Plastic invites a plasticity of forms but, perhaps, also an imperious indifference to "forms of ornamentation"; it rev-els "in sharp, clean lines and untroubled spaces."[8] In its vaunted double na-ture, plastic can also consort with ornamentation; this, for me, is plastic's other inherent countertextuality:

> Ornament may enhance the design of an object, when integrated with and subordinated to form. In the case of molded plastic objects, the possibilities for ornamental treatment are technically unlimited. The most intricate pat-

6. Meikle, "Into the Fourth Kingdom," 179.
7. Davis, "Life and Death in the Anthropocene," 350.
8. Gloag, "The Influence of Plastic," 467.

terns can be machined in the mold and readily reproduced. Nevertheless, a beautifully formed object, precisely made and of pleasing texture and color, generally needs little further embellishment. Purity and precision of form are important aspects of the appeal of machine-made products. Pure form provides its own ornament derived from the interplay of highlights and shadows, and the gradations of value created by reflected light. In a well-designed object, these provide pleasing, rhythmic patterns of light for the eye to explore.[9]

Plastic texture and, consequently, wit, bring forth a conflict between options of utility, economic viability, and bifurcation of taste. Interestingly, plastic welcomes its exploitation, its variegated manipulation, in keeping with the demands of the machine age. Plastic is never without its plasticity. It evokes the al*chemical* imagery:

Out of chemicals, powders, compounds, raw materials beyond the ken of ordinary folk, each of these objects had emerged with a single stroke of a hydraulic plunger. No clearly contrived series of human actions, indeed no human action at all, lay between conception and final form. Never mind that molds had to be designed, cut, and polished, or that moldings had to be finished and assembled/flash ground off, surfaces buffed, works installed, and so on. Their typical appearance was of seamless unitary creation.[10]

Plastic imports wonder, a mystic middle between appearance and reality and a deceptive simplicity.

Plastic has built its own "now"—the decisive advantage over its predecessors, the popularity of its present, and the messianic promise of a future. So plastic, as Eeva-Liisa Pelkonen notes, embodies "pure spontaneity without obvious intent."[11] She is right to observe that

---

9. Wallance, "Design in Plastics," 4.

10. Meikle, "Into the Fourth Kingdom," 173–74.

11. Pelkonen, "Plastic Imagination," 2. Pelkonen also writes, "A Paris-based group consisting of intellectuals, artists, and architects called Utopie, active in the late 1960s and early 1970s, deserves credit not only for actually using but also for articulating their attraction to 'synthetic' and 'ephemeral' materials like plastic in the magazine bearing the same name. That interest in new synthetic materials culminated in the 1968 exhibition Structures Gonflables (Musée d'Art Moderne de la Ville de Paris), which assembled various inflatable products, from high altitude weather balloons to beach balls, demonstrating in tangible ways how the material was shaping the aesthetics of the everyday, creating, in so doing, a new reality" (4).

unlike materials such as stone or clay, plastic does not exist in chunks of raw matter waiting to be worked with human hands. Indeed, plastic is in many ways unique in this regard: as a synthetic material, it retains its molecular proto-state and is able to metamorphose into any form without material friction. In the case of plastic, form and material are inseparable; rather than being imposed from without, form is generated by a single gesture, or what Barthes calls a "trace of a movement" within the matter.[12]

Plastic forms and deforms; it is a formation that builds the margin and involutely extends and rewrites the margins, flourishes on the possibility of construction and decenters itself through the prospects of travails and deconstruction. Within the material-aesthetic theory machine, the formation is the "plastic critical consciousness." Plastic is always "out of joint," though we do not ever realize it. It raises simulation, stylization, and a seduction that magnetizes people; plastic looks like wood or "credible imitations of marble" and projects an inadequacy that is attractive. Plastic affords and allows domestication; it helps the modernist imagination to prosper; it seduces modernist materializations. Plastic spells variety, producing new materials and moldings, different processing techniques (casting, lamination, compression), and different applications. Indeed, "Differences in chemistry, performance specifications, means of processing, function, appearance, value to consumers, and cultural meaning seemed of more relevance than any sense of unity bestowed by the near accident of convergence on the word 'plastic.'"[13] Plastic forms and spawns its own plasticity, evoking wonder, anxiety, fun, and often disgust, but also praise. The "plastic critical" is here to stay.

Plastics are, as Brennan Buck argues, "the materialization of virtual immateriality—without formal limitation or distinction between elements. Alternately hard and soft, stiff and malleable, plastics share a viscous continuity with digital form."[14] At the same time, polymer molecules and composite materials can also function "as discreet geometric assemblies with specific limitations and formal articulation."[15] The insight of seeing plastic with novel forms of tectonic expression may owe a debt to Gottfried Semper's

12. Pelkonen, "Plastic Imagination," 2.

13. Meikle, "Into the Fourth Kingdom," 174–175. See also Meikle, "Plastic, Material of a Thousand Uses."

14. Buck, "What Plastic Wants," 37.

15. Buck, "What Plastic Wants," 37. See also Mitchell, "Antitectonics," 206.

articulation of the difference between ground and frame and the significance that he attaches to the joint. Tectonics, pointing to the inner structure of a work of art and the "shaping and joining of form-elements to a unity,"[16] come to get highlighted in the work of Karl Botticher, Kenneth Frampton, and Eduard Sekhler.

Botticher's distinction between *Kerneform* (structural) and *Kunstform* (representational) contrasts with Semper's distinction between the technical and symbolic issues of manufacturing and the coherency that it might generate. In contrast to the idea of the "primitive hut," as expounded by Marc-Antoine Laugier in his *Essai sur l'architecture*, Semper introduces the notion of the "nomadic hut" and, with it, the significance of the knot and joint in the tectonic process.[17] Knots are symbolic of expression, and joints grow a complicated poetics of construction. Keeping the importance of knots in focus, Semper sees knotting as a means of expression, as a tangle and a plexus of forces that "joins and commands everything." He argues that "the sacred knot is chaos itself: a complex, elaborate, self-devouring tangle of serpents."[18] This knot is contextualized in acts of "weaving, netting, knitting, braiding, and embroidery, crafts that render the knot indistinguishable from its context, integral to a larger network."[19] Plastic manifests within such tectonic knots that are entangled forms of connection building expressive multiscalarity through structural diversities, aesthetic experiences, and cultural appropriations; knotty plastic sways between the symbolic, semiotic, synthetic, and the singular.

Another significant material-aesthetic dimension of the plastic turn is transparency, a category that has potentially ambiguous ramifications in the aesthetic imaginary of the twentieth century.[20] Transparent plastic— packaging and design that create entirely new products such as clear "blisters" providing unobstructed vision for airplane pilots and light-piping devices for medicine, dentistry, and display design—exerts some advantage over

---

16. Andersson and Kirkegaard, "A Discussion of the Term Digital Tectonics," 30.

17. See Semper, "The Four Elements of Architecture"; see also Frampton, "Rappel à l'ordre."

18. Buck, "What Plastic Wants," 38.

19. Buck, "What Plastic Wants," 40.

20. See my "Aesthetic Imaginary." "Mostly undetermined, indeterminate, and capacious, aesthetic imaginaries aggregate around dwellings in culture, social practices, characters of imaginative reconstruction, and affiliations with religious and spiritual denominations and preferences. The aesthetic imaginary is built inside the borders of a nation, a culture, a society, a tradition, or an inheritance; but, it disaggregates and reconstructs itself when exposed to the callings and constraints of cross-border epistemic and cultural circulations" (450).

glass but both mediums see the "other side" differently and encounter the reality of *seeing* differently as well. The "looking through" is different. Glass usually breaks, plastics can be relatively stronger; plastics are lighter, glass heavier. Plastics and glass acquire a separate aesthetic and communities of taste. Plastics contaminate; glass are more immune to diffusive pollution. Plastic is imagination; glass is closer to reason. Plastic is also the more restless of the two, morphing, adjusting, experimenting, and molding itself to fit and cut shapes and spaces. Perhaps that shows through in the innovation of bubbled plastic print (as Su Liehsi points out): it contributes to a new form of printmaking.[21] Here the latent countertextuality is hard to evade, as David Ruy observes:

> Recently, it has been interesting to see the narrative of PLA [polylactic acid] plastics being incorporated into 3D printing technologies. The old magic of seeing plastic pellets being formed into bottles by machines is getting a refresh through the magic of extruded plastic filaments being melted and deposited in micro layers by robots. We're told that PLA is made not from bad petroleum but from nice biomass (starches from corn, potatoes, or beets). Though this story conjures an image of a rehabilitated plastic returning gently back to the soil (not unlike a corpse), the reality is that PLA is also quite resistant to losing its form (not unlike a soul). Most of it will still end up in the landfill, where it will be stored for some unknown future. As these awful circumstances accumulate, at some moment you may notice that *form does not enter or exit from nameless matter so easily*. Whether it's plastic, metal, or even flesh, *nameless matter* is not so easily manufactured. Form has to be forced onto matter, and even more force is required to remove it. Whether it is a literal forcing via grinding, shredding, or incinerating machines or a conceptual one via ontologies of pure becoming, objects tend to stay objects until they turn into other objects when the human being is not there.[22]

Despite efforts to recycle, compost, and incinerate, the adamancy, perhaps, the irreducibility of plastic—a somewhat nameless, auto-creative

---

21. Liehsi, "The Bubbled Plastic Print." Immediately after bubbled plastic print was developed, it aroused great interest in art circles both in China and abroad. Bubbled plastic print also gives full play to the artist's spontaneity. After seeing some bubbled plastic prints, Mr. Liuhon, a Japanese expert in graphic art, said in excitement: "No other form of graphic art can exert such artistic charm."

22. Ruy, "Review of Pelkonen, 'Plastic Imagination,'" 7 (emphasis added).

form-ability—graduates to unimaginable forms. It keeps recentering its own degradability. Disintegration and dissolution and dissipation are forms of becoming plastic; it is material imagination. Plastic builds its own plastic, which, like textualism, is plasticism. How inherently *countertextual* is plastic? It builds enthusiasm to create, but the enthusiasm translates to "de-creation" as well. If there was wonder in the way it grew its own trajectory, the wonder today registers on a note of recalcitrance and hegemony over the environment at large. Efforts to dislodge, degrade, and decry plastic have met with "toxic" resistance, for plastic continues to contaminate. Plastic is bioresistant, and yet efforts persist to biodegrade it. For instance, Nylon 4 degrades through contact with the microorganisms ND-10 and ND-11; polyurethane degrades through exposure to several species of soil fungi; and the fungus *Aspergillus fumigates* breaks down plasticized PVC. Plastic resists, plastic succumbs. This countertextuality of plastic that the turn has helped us to see or has brought upon us proves unprecedented as it changes the way we think, act, express, and exist.

The American painter Mark Rothko sees plastic creativity as being more in line with the antinomies of romantic creativity. He argues that "by the constant rearrangement" of its properties, "art, like every other species, proceeds according to logic through stages of change that we can call growth. It grows logically, definitely, step by step from the exhibition of one set of characteristics to another, always related to its past equipment, and bearing at the same time the promise of the future."[23] Emphasizing plastic laws, Rothko argues for its ability to preserve "a continuous, logical, and explicable picture" for art; such laws contribute to the "organic continuity of art." Plastic submits to the processes of objectification; it is "organic" and logical, but not without its own "imagination." This is not the eighteenth-century plastic imagination within the framework of German romanticism or the relevant writings of William Duff, Alexander Gerard, and the Earl of Shaftesbury. Keeping the notion of nonidentity in play, plastic has combined pliability and hardness—a seeming suppleness and an obstinacy that refuses to change or is resistant to change; it is welcoming and a barrier, embracing and again a sealant, generative and, again, deeply obstructive to its degeneration. In fact, twentieth-century thought is informed by this plastic-countertextuality. Plastic embeds plastic: it is an event

---

23. Rothko, *The Artist's Reality*, 14.

that is resilient until the breaking point; melts and morphs under heat; is enzymatically dynamic; can be explosive; is a protective cover for the characteristic fragility of glass, and a monotony in industrial self-sameness. The aesthetic imaginaries built around plastic are enticed by the power of the banal, the consumerist tendencies to succumb to the plastic-pressure (call it the plastic-popular), the bounds of inventiveness within the hard limitations of producing plastic, the contrastive malleability of creation as against clay or clod, the complicated dialectic between freedom and unfreedom and the conjugation of the Arts and Sciences. The plastic imagination, as propounded in eighteenth-century aesthetics, is a constructive force with a proto-romantic strength of image-making. With the turn, the plastic imagination is formed in the trans-dialectic—the anxiety, the con, and the contrary—which points to a complexity that is creative. This is not because it produces a new being or form; it is because of a creativity that countertextualizes in the helplessness of submission and the fragility of a surrender, the urge to avoid and the compulsion to engage—the love-problematic enmeshed in seduction and revulsion. The negative dialectics works best through such plastic thinking. Is plastic hard or malleable? Is critical thinking hard or soft?

Material-Aesthetic II

Whether it is Goethe's *Elective Affinities* that brings chemical metaphor to bear on human relationships;[24] or Primo Levi's *The Periodic Table*, where the metaphoric articulation through zinc, lead, mercury, carbon, and others seeks a relationship of chemistry to the value and beauty of impurity (as prescribed in Nazi theory)[25] and to Jewishness, race, and identity;[26] or Mala Radhakrishnan's chemistry poetry collections, *Atomic Romances, Molecular Dances* and *Thinking, Periodically: Poetic Life Notions in Brownian Motion*, where atoms and molecules connect and engage with subjects ranging from

---

24. McKinnon, "Elective Affinities of the Protestant Ethic."

25. Wilson, "Primo Levi's Hybrid Texts."

26. Levi, *The Periodic Table*, 34. Using zinc as metaphor, Levi observes: "I am Jewish. . . . I am the impurity that makes the zinc react, I am the grain of salt or mustard. Impurity, certainly, since just during those months the publication of the magazine *Defense of the Race* had begun, and there was much talk about purity and I had to be proud of being impure." See also Baumgarten, "Primo Levi's 'Small Differences' and the Art of *The Periodic Table*."

finding love and making friends to pursuing dreams; or Joanna Buckley's collaborative poetry project with the University of Sheffield, *Periodic Table of Poetry*;[27] or Roald Hoffmann's "chemist" poems as collected in *The Metamict State* and *Gaps and Verges*[28]—we are in the midst of chemo-poetics of different orders that connect life and its undulations with molecular formations. In *Polymers*, the Canadian poet Adam Dickinson, sees polymers and poetry in a bind and bond that bring writing and chemical models into a dialogue, a kind of fabrication in which every poem in the collection speaks and represents a particular chemical structure and process. Further, the poems' titles in *Polymers* all begin with the letters C, Cl, H, or O, the elements that comprise the most common plastic resins.[29] The materialist poetics here, as also in Christian Bök's *Crystallography*, is conscious art: art in design and designated with a clear agenda to achieve and execute.

Dickinson, intrigued by the "overlap between the synthetic and the natural, between writing and rewriting, between the metaphorical and the literal, and between the god-like creation of new matter and the hellish consequences of things that refuse to decompose," sees "polymers all over the place in language and rhetoric (anaphora, polysyndeton, puns), as well as in social and cultural behaviors (fashions, memes, obsessions)" and devises a project where poetry is imagined to perform "an experiment to expose these cultural polymers"; through the poems he aims to "re-imagine the structure of polymers (their chain-like dynamics) in terms of their cultural and linguistic analogues."[30] As the critic Treasa De Loughry notes, he writes "multi-scalar poems that imaginatively repurpose chemical structures to produce new poetic principles"; his collection is "a poetic-chemical way to analogize chemistry, language and society."[31]

Dickinson's "petro-poetics" in *Polymers* is partly in line with what I call the material-aesthetic. He consciously crafts his poems in correspondence with polymeric structures in a kind of novel execution of poetic experimentation in

---

27. "Periodic Table of Poetry," accessed September 26, 2021, https://periodictable.group.shef .ac.uk/.

28. Hoffmann, "The Poetry of Molecules," 142. It is interesting to follow the way Hoffmann talks about hemoglobin and beauty ("the way this molecule travels through the blood vessels while constantly transforming is as thrilling to me as the story of Odysseus").

29. Dickinson, "In Conversation."

30. Dickinson, "In Conversation."

31. De Loughry, "Polymeric Chains and Petrolic Imaginaries."

form and expression. However, within what I characterize as the "Plastic Literary" in chapter 2, the plastic-polymer texts (which is how I qualify them) are not consciously crafted extensions of chemical formulas. If Dickinson advocates a deliberate extrapolation of literary-cultural space through plastic behavior, my argument is fundamentally about the "plastic material" as deeply embedded in art and letters. Dickinson or Radhakrishnan *figure* their poems with the consciousness of polymeric or chemic properties. But the material-aesthetic in *The Plastic Turn* has an "unaware," rather, post-configurative aspect to it in that *The Waste Land*, for instance, was not conceived with any polymeric consciousness, but becomes "polymeric" as part of a plastic literary—a deep plasticity. We have always been plastic!

Plastics are organic polymers with high molecular mass, which make them composite substances. This helps plastics to have inter-substantial presence, imputing a low decomposition rate. Plastic has a combinatory power and plexus: the ability to incorporate, constellate, and aggregate, and, again, exhibit resistance to biodegradation, perishing, and ruin. In fact the chains of carbon atoms become organic through the presence of oxygen, nitrogen, and other elements; there are "repeat units," and polymer chains may have numerous repeats units that form their own points of linkages.[32] It is interesting to note that with "side chains" in place (see Figure 1), plastics can also be available in such forms as acrylics, polyesters, silicones, polyurethanes, and halogenated plastics, and the chemical processes employed in their synthesis include condensation, polyaddition and cross-linking.

Plastics survive through a variety of physical properties: they have hardness, density, tensile strength, and resistance to heat, combining and exuding variety. The term *plastics* applies to a wide range of materials that at some stage in manufacture are capable of *flow* such that they can be extruded, molded, cast, spun or applied as a coating.[33] Plastics can be crystalline, semi-crystalline, and amorphous in their molecular structure. This results in a range of glass transition temperatures—the point at which the material's structure shifts from rigid to rubbery—and flow-tensions created by intermolecular forces affecting the tensile strength and flexibility of the polymer.

---

32. For more on plastic structure and polymerist properties instrumental in understanding what I call the plastic intertextuality, see Harper, *Handbook of Plastic Processes*, and Birley, Heath, and Scott, *Plastic Materials*.

33. Thompson et al., "Our Plastic Age."

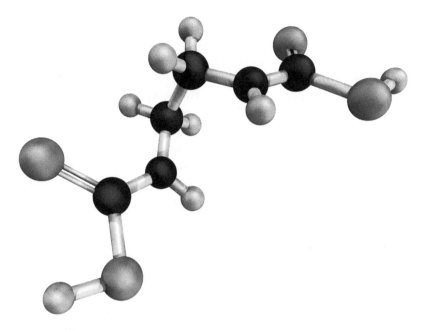

Figure 1. Three-dimensional polymer structure. Source: Shutterstock.

So plastic accumulates, aggregates, intertextualizes, dissipates and disintegrates.

The material *intertextuality* of plastic corresponds aesthetically with the modernist appropriation of cross-linkages and cross-overs, spilling over into a number of thought-experiments across continents and resonating through the entire epistemic swathe of the twentieth century. The plastic literary within the framework of such an understanding, as exemplified in the next chapter, informs the turbulent critical plateau of twentieth-century consciousness across world literature. So questions such as—To what extent is Gertrude Stein plastic? How plastic is Steinese? How "additive" is Anti-poetry, the work of Takechi Tetsuji, and the Jikken Kōbō collective? What makes Eliot's *The Waste Land* a PVC (polyvinyl chloride)? What makes Umberto Eco's novels instances of plasticizer-textuality? Is Hélio Oiticica's art semi-crystalline in nature?— become deeply relevant to how we understand modernism and postmodernism.

Plastic's relationship and collaboration with additives make for an interesting material-aesthetic negotiation. Additives like stabilizers, fillers (cellulose, wood flour, ivory dust, zinc oxide), plasticizers (adipates and phthalates) make plastic more controversial and complicated in its functioning and relationship with nature. They make plastic toxic, degradation-resistant, flame-retardant, lighter, and more viscous. They alter and vary the properties and behavior of the final plastic. The industrial use of additives corresponds with the conceptual position around hyperallusivity and creative appropriations in modernist and postmodernist poetics, enabling a potential material-aesthetic to unfold. The plastic turn inspires a sustained skepticism and romance to include or *add* something to the existent, the given, and the happening. This "additive-urge"—deeply integral to the critical consciousness of the twentieth century—is inexorable as it continuously invests writing and thinking in a series of innovations and movements. Through its trans-formations, plastic brings "entextualization" and decontextualization through a variety of co-texting with other materials: the kind of textualization that intertextuality as a strategy and process constructs for itself. Richard Bauman observes that

> the process of entextualization, by bounding off a stretch of discourse from its co-text, endowing it with cohesive formal properties, and (often, but not necessarily) rendering it internally coherent, serves to objectify it as a discrete textual unit that can be referred to, described, named, displayed, cited, and otherwise treated as an object. Importantly, this process of objectification also serves to render a text extractable from its context of production. A text, then, from this vantage point, is discourse rendered decontextualizable: entextualization potentiates decontextualization. But decontextualization from one context must involve recontextualization in another, which is to recognize the potential for texts to circulate, to be spoken again in another context. The iterability of texts, then, constitutes one of the most powerful bases for the potentiation and production of intertextuality.[34]

The plastic manifestation and its inherent structural status point to further modernist substantiations. Bakelite is "composite" in an entextualization as varied as phenol and formaldehyde with wood flour, asbestos, and slate dust. Polystyrene or "foamed plastics" demonstrate an "open cell" form of interconnected foam bubbles. Polyvinyl chloride (PVC) with C-CI bonds can resist oxidation and grow strong and weather resistant; it can also be softened with chemical processing. What we call plastic leaching is the material-aesthetic cognate for "conceptual leaching." Plastic breaks borders of appro-

---

34. Bauman, *A World of Others' Words*, 4.

priation, both leaking *out of* itself and leaking *into* something else. It transcends itself through associationism, intersubjective overlap, recurrence and play. Plastic is never plastic if it cannot be *with* "others." A text is never outside intertextuality. Once plastic, always plastic; once text, always intertext.

Moving past celluloid and Bakelite, plastic forms its own emergence with stunningly expressive outlines. *Synthetica*, the map of a plastic continent with such countries and regions as Melamine and Lignin, published in *Fortune* magazine in 1940, spoke to the spread and reach of the material: a plastic map and mapping plastic combined together with a politics and aesthetics of their own. Contrasting *Synthetica* with *Fortune*'s "An American Dream of Venus" photo feature, Meikle observes that "unlike the map of *Synthetica*, which postulated a land mass ripe for systematic exploitation, the plastic Venus promised an American dream of shifting shapes, an irrational phantasmagoria of ungrounded images, all in brilliant synthetic colors, a carnival of material desire."[35] Plastic promises abundance: "a world in which man, like a magician, makes what he wants for almost every need, out of what is beneath him and around him, coal, water, and air." This magicality and immensity build an aura: the kind of excitement evident in Edwin E. Slosson's rapturous passage (in 1919's *Creative Chemistry*) about "Man the Artifex," molding a world "nearer to his heart's desire."[36] As Bernadette Bensaude-Vincent points out, "matter came to be presented as a malleable and docile partner of creation—a kind of Play-Doh in the hands of the clever designer who informs matter with intelligence and intentionality. Just like the *demiurgos* in Plato's *Timaeus*, the material engineer can impose forms on a passive, malleable *chora*."[37] This brings us before what I call "plastic avant-gardism."

Plastic's material miscellany (polyamides, polycarbonate, polyethylene, polypropylene, polyurethanes, polyvinyl chloride, etc.) is avant-gardish, always escaping a unifying and univocal image. Meikle points out that

> the materials were too distinct to lump them together. Hard, dark, infusible phenolic bore little resemblance to bright, colorful, thermoplastic cellulose acetate, or to soft, sometimes tacky vinyl sheeting, or to hard, glasslike formable acrylic. Different chemistries—those of cellulose, coal tar, and petroleum—divided the

35. Meikle, *American Plastic*, 67.
36. Meikle, *American Plastic*, 69.
37. Bensaude-Vincent, "Plastics, Materials and Dreams of Dematerialization," 22.

world of plastic. Different *processing techniques*—casting, laminating, compression molding, injection molding, extrusion of thermoplastics like vinyl, machining of cast phenolics and acrylics—came from several industrial traditions. *Different applications* revealed to a range of end-use industries—from radios and automobiles to toys, tioletware, furniture, and housewares—each with its own needs, preferences, and ways of doing business.[38]

In line with plastic and its plasticity (the applicative and process-technique dimensions), we encounter in the avant-garde "the dispersion of subjectivity and the crisis of artistic expression, the differentiation of life spheres and the relativization of structures of knowledge, the instability of the real and the contingency of the ethical."[39] With the turn, plastic and modernism concomitantly left us in the "folds of interlocking political, economic, social, and technological developments."[40] If Europe was changing its image after the First World War—working to reframe, in Paul Valéry's formulation, its "crisis of the spirit" into a "new Europe"—plastic changed our way of looking at Europe too. *Synthetica* outlined our plastic borders and underlined our spatialities of engagement with various modes of aesthetic-cultural expression. The old Europe had met with its limits and so did our material politics and practicalities with wood, ivory and the rest. Plastic "outbounded" thinking and brought a new direction to ways of mattering with matter and, consequently, subjectivities.

With the turn, plastic takes liberties with use and application in a peculiarly modernist way: it is synthetic, artificial, assimilative, aggregative, and co-constructive. If progressive poetics challenge the expectations of the reader, plastic challenges the consumer. It is indeed difficult for both the reader and the consumer to comfortably position themselves with the objects resulting in a disquiet that declares no chaos but an imaginative reconstruction. What this does is to destructure and dramatize the already existent or the readily received, projecting a different ego of the work-aesthetic and the material. This expressionist revolution (as with the material and the material-aesthetic in the plastic literary) is not reckless and never without a conscious intent and technique. The flourishes of plastic expression cannot disconnect themselves from cultural fashion and the aesthetic and institutional demands of art and

---

38. Meikle, *American Plastic*, 98 (emphasis added).
39. McBride, McCormick, and Žagar, *Legacies of Modernism*, 4.
40. McBride, McCormick, and Žagar, *Legacies of Modernism*, 4.

utility. Revolution is simultaneously radical and, in its sense of a "return," (re)conservative: the plastic turn achieves both artistic reconstructionism and a change in ways of thinking or a view of life. Working outside its own normativity—organically and constitutively—it spatializes the culture of aesthetic consumption as people start to become accepting of hybrid forms that go beyond the usual partnership of wood and glass. Has plastic changed our point of view? Has it taught us a new philosophy of relevance?

There is a category of high-performance plastics. Significantly, the word *performance* premises the impact and efficacy of their execution. Plastic performance, modernist performativity, and post-modernist avant-gardish high performances share points of analogical linkages. Plastic, however, does not herald arrière-gardes in all exclusivity for similar to modernist unfoldment and enfoldment it questions its existence against materials it competes and, mostly, conquers. Its arrière-gardes hold a paradox—similar to what modernism holds in its unfurlment through the first half of the twentieth century—which produces a complexity of emotion, attitude and ways of engagement. Plastic over-experiments and, continually, unsatisfied with its latest organic incarnations, chooses to reemerge in fresh formats and chemistry. It triggers enthusiasm in use and reach but loses class and pedigree, has built its ubiquitous power but comes to be tagged as "plastic" and denotatively downgraded as classless, unexciting, prosaic, and plebian. Plastic has marched along, but not without the sticking burrs of cultural animadversion and an increasing concern of its pollution-potential, getting a somewhat elitist and classy stare from wood and ivory and other rare metallic consumerist versions. This attests to another version of its countertextuality.

Roland Barthes introduces the idea of plastic's (un)ease, antitheticality, and somewhat dubious identity by noting that although it has variations as having names like Greek shepherds (Polystyrene, Polyvinyl, Polyethylene) but is "in essence the stuff of alchemy." It becomes an event in "transmutation of matter": "more than a substance, plastic is the very idea of its infinite transformation; as its everyday name indicates, it is ubiquity made visible. And it is this, in fact, which makes it a miraculous substance: a miracle is always a sudden transformation of nature. Plastic remains impregnated throughout with this wonder: it is less a thing than the trace of a movement."[41] Its proliferation in forms and

---

41. Barthes, "Plastic," 92.

plurality of effects project "amazement" manifested through pleasure and power—"the euphoria of a prestigious free-wheeling through Nature." But plastic for Barthes also has its disenchanting moments—the principle of countertextuality—in that it is seen as a "disgraced material" losing out on its hierarchical superiority over the "effusiveness of rubber" and the "flat hardness of metal." The first (rubber) keeps you in a state of mind that anchors without the embeddings of a confident pleasure and the second (metal) establishes a settlement in understanding and security of thought; the former gives exhilaration, the latter joy; plastic provides the unease of a potential disequilibrium as against metal's contentment and poise. Plastic is "powerless to achieve the triumphant smoothness of Nature" and is imitation, and, hence, pretentious. But its pretentiousness is infectious for it invades the household space with an inevitability and insurgence, reappropriates the principles of usage, and repremises the content and the need of the society—"the first magical substance which consents to be prosaic."[42] Does it reinvent our negotiation with the materiality of matter, proclaiming no triumphant reason for its existence without being silently all pervading? Can we figure out what plastic can do to us? Or is it both memory and immemorialization, a cognitive conceptual representation and beyond?

The exuberance is not without its own problems, spreading "half-truths, essentially misleading tidbits,"[43] and the illusion of plastic being able to achieve the impossible and the every beat of the heart's desire. This can be seen as an interesting development in the complexities of the plastic turn, something that I would like to characterize as the "counter." If the principle of the modernist counter sees enthusiasm as riddled by doubt, exuberance being made wary of straying into delusion, and experiments being overlaid by possibility and implementation but not always without discretion, plastic counters rubber and glass, and it (en)counters a massive success in its counter-emergence. Plastic counters because it is inherently counter-poetic: habit to helplessness, seduction to submission, discovery to dependence, freedom to incarceration. Also plastic today has a different politics of counter which includes countering aqua-biotic balance, "genetic soup," revulsive and ignorant humans and hapless nonhumans. Sy Taffel argues that "certain forms of bacteria have evolved to reside within marine plastics, forming a microscopic ecosystem that has been referred to as the plastisphere: 'a diverse

---

42. Barthes, "Plastic," 93.
43. Meikle, *American Plastic*, 73.

microbial community of heterotrophs, autotrophs, predators, and symbionts.' These organisms, whose continued existence requires the presence of marine plastics, raises questions about how we assign value to different life-forms and further troubles straightforward proclamations that plastic is pollution."[44]

Plastic contraposes through the interspecies mesh—the "charismatic marine megafauna such as turtles and whales" with "alien microbial life-forms who dwell within oceanic plastic gyres."[45] Material-aesthetically, twentieth century critical consciousness thrives on counters: Samuel Beckett contra Aristotle, James Joyce contra Arnold Bennett, T. S. Eliot contra the Romantics, Heidegger contra Greeks, Lyotard contra Kant, Hayden White contra Ranke, Orientalism contra colonial discourse, as the plasticity of the list can be heavily restive. Also, the materializaton and the material manifestation of plastic have an interesting dialectic to unravel: plastic forms through a processual complexity uncommon in other material counterparts but its manifestations are mostly smooth, organic, harmonic, and usually unified; for instance, furniture, toys, and various other incarnations. However, plastic's manifest forms conceal their complicated processes of formations: a kind of self-countering and disruption, as it were. This is, often, very modernist: a kind of progressive poetics that speaks of non-organicity and discontinuity.

In a modernist paradox formed around the status of the present evolving out of avant-gardism and tradition, we encounter a crisis of a different order. Modernism has struggled to premise the concept of literature, has wrestled with a certain devaluation brought about by over-experimentation, and agonized over the necessity to open up the literary and an embattled consciousness that exceed its own making for further recesses of meaning and signification. Much in the same way, plastic struggles to define its contribution that is never without its share of ambivalence. The present for both plastic and modernism is vexed and restless. Plastic modernism is antithetical: the tension to define itself is the paradox of living in a crisis. A victim of modernist restlessness, plastic introduces creative intertextuality, a fold in time. The modernist timber of thinking is surely the plastic *con* affect.

Plastic modernity has brought, in the words of Andreas Huyssen, "modernism at large"[46]: it has taken its own time, evolved through complicated

---

44. Taffel, "Technofossils of the Anthropocene," 361.
45. Taffel, "Technofossils of the Anthropocene," 361.
46. Huyssen, "Modernism at Large," 57.

matrices of power, capital, economy flow, representation, and politics of art and material manifestations. Like modernist texts and their critical appropriations, plastic builds its own politics and logic of migration, gets dominated by market forces, strategies of distribution, and post-national curiosity and commercial accessibility. The question is whether plastic challenges the high/low distinction, the hierarchical and vertical value scale of consumption, public taste, and other aesthetic configurations. Plastic's emergence, strategies, and strata of distribution complicate the value structure of the culture industry and the culture of consumption. In contrast to wood, which tries to preserve its high status, plastic finds itself relegated to the low and indexed economically through availability and affordability. But the practices of use and innovation bring plastic to break down such distinctions and constellate with wood in the bricolaging of "arty" objects that, on most occasions, have wider consumerist attention and acceptance. So in its discovery and consequent evolution and transformation, plastic does not imply a radical and wholescale repudiation of the past; indeed, it has a past with other materials (for instance, coal, natural gas, cellulose, salt, crude oil) that contributed to the material emergence of plastic.

Like the modernist notion of tradition, plastic is disruptive and oppositional but not without a sense of continuity. Its constitution is made possible through elements that precede plastic: these are the pasts that live in deep intertextuality to see the emergence of the plastic-present. Plastic metonymizes self-transcendence and mobility, and it keeps transforming its own tradition and evolution out of a mix with and in other substances and processes. Plastic evolved out of itself in a relentless urge to index its limits and achievements. Modernism is "not a univocal movement"[47]; moderns have "antimodernes" (in the words of Antoine Compagnon[48])—much in the same way plastic is not constitutively atomistic. Both are steeped in paradoxes and have inherent formative tensions. Plastic gives a new definition of the present: it places itself at the cross-roads of a past which does not have plastic and a future which cannot think outside plastic.

---

47. Marx, "The 20th Century," 66.
48. See Compagnon, *Les cinq paradoxes de la modernité*.

Material-Aesthetic III

With the plastic turn we are in the midst of a reformulated "poetics of relation," transforming how we think and experience our socioeconomic realities,
existential questions related to our life and fragility on Earth, struggle for
survival, geomorphologies, precarity and disposability. Plastic has always had
hurry in its production and a flair and flamboyance in its presence in the
world at large. However, it is "slow" in many other forms of being and motion, whether in use or after-use or long after it had gone out of use. It has a
"slow kinesis" that violates our physiological barriers and renders "delayed
destruction" that is "incremental and accretive, its calamitous repercussions
playing out across a range of temporal scales."[49] If plastic's slow kinesis has
brought representational, existential, and biopolitical changes, our well-
meaning patterns of thinking and thoughtful execution have changed as well.
Although plastic has brought all in a commonly inhabited space of use and
abuse, it has increasingly produced a divide among people of different classes
and communities. Geopolitically, people and plastic are not uniformly connected as poverty, unemployment, the commerce of recycling, "trans-boundary
pollution,"[50] unhealthy living conditions, and various forms of deprivation
create segregated zones of livelihood and compulsive settlement. We cannot
ignore separate plastic-fates as scales of impact and the degrees of disaster
vary. This produces "dissonant communities" as distributive plastic-use paradoxically produces plastic-discrimination—plastic as the eco-material—and
becomes a pathological marker to determine the plastic-haves and less-plastic-
haves in an equation where there are no plastic-have-nots.

Plastic-drift and seepage are close to being cosmopolitan and secular in
nature, cutting across all biotic and nonbiotic forms—lethal and permeative,
perverse and pervading. Plastic has experimented transgenerically (as chapter 2 elaborately demonstrates) and has also contaminated transcorporeally.
Plastic adds, absorbs, releases, effuses, transfers, and accretes; it is combative
and cooptative. The issue of plastic-connect is far more encompassing than we
thought it was or has been. If turtles feed on plastic bags mistaking them for

---

49. Nixon, *Slow Violence and the Environmentalism of the Poor*, 4.
50. Whitehead, *Environmental Transformations*, 9.

jellyfish, "plankton trawls recover substantial quantities of plastic."[51] It is argued that "several persistent organic pollutants (POPs) bind to plastic as it is transported throughout a watershed, buried in sediment, or floating in the ocean. A single pellet may attract up to one million times the concentration of some pollutants as the ambient seawater, making those chemicals readily available to marine life. Food mimicry, based on color, shape or presence of biofilms, is one mechanism driving wildlife to ingest plastics, in addition to filter feeding and respiration."[52] Plastics not only have the potential to transport contaminants but they may also increase their environmental persistence: "The situation for higher-trophic-level organisms is more complex because of 'biomagnification.' Tissue concentrations of hydrophobic and poorly metabolizable contaminants, such as PCBs, are amplified through the food web. Higher-trophic-level organisms (e.g. seabirds) are exposed to highly enriched concentrations of hydrophobic contaminants via their prey (e.g. fish). Therefore, ingested plastics (i.e. anthropogenic prey) compete with the natural prey in terms of contaminant burden to the predator."[53] Stacy Alaimo argues:

> Although the recognition of trans-corporeality begins with human bodies in their environments, tracing substantial interchanges reveals the permeability of the human, *dissolving the outline of the subject*. Trans-corporeality is indebted to Judith Butler's conception of the subject as immersed within a matrix of discursive systems, but it transforms that model, insisting that the subject cannot be separated from networks of intra-active material agencies (Karen Barad) and thus cannot ignore the disturbing epistemological quandaries of risk society (Ulrich Beck). As a critical posthumanism, trans-corporeality, by insisting on the material inter- and intra-connections between living creatures and the substances and forces of the world, denies human exceptionalism by considering *all species as intermeshed with particular places and larger, perhaps untraceable currents*.[54]

Plastic dissemination and contamination encourage and trigger a trans-corporeality whereby the fish, the bird, and the human come to be enjoined

---

51. Teuten et al., "Transport and Release of Chemicals from Plastics to the Environment and to Wildlife," 6.

52. Eriksen, "The Plastisphere," 6.

53. Teuten et al., "Transport and Release of Chemicals," 8.

54. Alaimo, *Exposed*, 132 (emphasis added).

both inter- and intra-actively. Being in land, sea, and air is being in plastic, in both traceable and untraceable currents. If plastic connects, plastic extends; if plastic permeates, plastic binds; if plastic spawns, plastic trans-corporealizes. Our refusal of plastic is, paradoxically, an entanglement in plastic; plastic is at once our prior, intermeshed reality and our becoming. What I mean is that a plastic use–free zone is entangled within a plastic zone; plastic is the interminable center that, once removed, discovers its other centers of operation and manifestation. Within such plasticities, plastic has brought us together—human and the nonhuman—in life and death, in our plastic-sympathy and plastic-ethics, prompting, ironically, our bio-egalitarianism a rethink. Plastic's impact on our ethical ways of thinking the other—the "ethical imagination" in general—has a meaningful correspondence with how literary works have formed their "post-literary" since the days of high modernism. This concept of "plastic transcorporeality" is clearly evident in modernist literature where the "literary" allows generic and paradigmatic transfusion, creates spaces for transdisciplinary inroads and expropriations, and makes thinking a shared habitation across faculties of human knowledge. Plastic has shown that form and content vary in continual correspondence with structural manifestations. This comes to review the ways in which we understand the "literary." The "literary" is here and there, contained and released, unformed and forming, resistant and overtaken, in and out of the discipline, romantic and imitative, degraded and nondegradable. If plastic has redefined our notions of connect through its (in)visibility of collaboration and collusion, modernist and postmodernist literatures are no stranger to it. For instance, Virginia Woolf's fiction reformulates the "literary" through sciences, quantum mechanics, entomology, readings in the unconscious, and a variety of other epistemological persuasions. Plastic's lethal and lissome ability to connect all—a reconceived Darwinian "tangled island"—spells out the delicate and delectable power of twentieth-century literature to push forward its own agenda of "dwelling," a provocative cosmopolitanism of existence and its commitment to "migrant reading."[55]

Plastic is the reality between utility and pollution, between the popularity of its use and its nonbiodegradable menace, between utilitarian comfort

---

55. For more on "migrant reading" see my *Trans(in)fusion*.

and the earth as trash. The plastisphere appeals and appalls. It is wide-ranging in its reach, impact, and consequence:

> Through degradation by sunlight, biodegradation, chemical and mechanical degradation, plastics fragments disperse globally, accumulating in massive circular currents called subtropical gyres, where wind and waves slow down toward the centers. Microplastics less than 5 mm to macroplastics of all sizes above have been reported since the early 1970s in the subtropical gyres of the North Atlantic, South Atlantic, North Pacific, South Pacific, and outside the gyres in nearshore environments. They have also been found in estuaries, lakes, closed gulfs, bays, and seas. On land, plastics dominate desert landscapes, and wind-driven micro and nanoplastic particles can reach distant terrestrial biomes, evidenced by the inadvertent collection of these particles by pollinating insects.[56]

Hence, plastics disseminate their *presence*. Jody A. Roberts notes that "the spread of plastic has been more subtle, and it is perhaps for that reason that experts of all stripes missed it slipping into unintended places, travelling near and far such that nearly every cup of water from the ocean is likely to contain some plastic in some form of degradation and nearly every human subject found anywhere on the globe will likely bear the marks of a plastic modernity."[57] Plastic dissolves the divide between the local and the global, trans(in)fusionizing its presence,[58] such that every drop of ocean water and every human being *across* race, culture, political denomination, and religious affiliation have some lasting "plastic" in him or her. It de-divinizes the local-global bind to evolve in ways where the "literary" *potenza* (power, strength, and potency) is in being-with and being-across. This approximates what Susan Friedman calls "planetary modernism,"[59] where local formations are deeply inflected by cultural traffic of all kinds. Plastic disseminates and all localization of plastic is immanentization of plastic: plastic used here is plastic experienced elsewhere, (t)here. Plastic parataxis hybridizes to a point that

---

56. Eriksen, "The Plastisphere," 2–3.

57. Roberts, "Reflections of an Unrepentant Plastiphobe," 111.

58. The directive power of plastic comes through my trans(in)fusion ways of doing literature and humanistic thinking: the regulative, dispensing, projective, proleptic, trending spirit that substantializes itself on cultural and epistemic border-crossings. How much of *trans-now* and "taking place" is plastic? See Ghosh and Miller, *Thinking Literature across Continents*, and Ghosh, *Trans(in)fusion*.

59. See Friedman, *Planetary Modernism*.

its formations overlap a variety of things, and its presence is unfixed by geographical and sociocultural borders around the world; it is "permeable," "porous," and pervasive. This material aestheticizes the twentieth-century turn in thinking literature *across* continents. I call this "plastic literature."[60]

Tradition and change, passion and propriety, experiment and heritage are involved in a subtlety that is integrally plastic in nature. Plastic, ever since it burst into the scene with color and passion, has never allowed us to settle in a way that is devoid of anxiety and aspiration. Plastic unrest and unease have become the permanent fate of the arts and letters. The plastic turn makes the plastic-global become the new reality—the translocational momentum and might that do not discriminate between their point of origin and eventual destination, between race, ethnicities, discriminations of color, religious denominations, cultural heritage, and political schoolings. Plastic-mondialization diversifies the transnational spaces and cross-border appropriations and repremises the local-global divide in terms that are far more complicated than what conventional understanding would allow. We are inevitably in the midst of plastic-worlding.

The plastic turn, hence, has forced us to rearticulate the philosophy of the Other: communities of conversation and exchanges, the principle of I-You, and the "literary Other." We live in a world of plastic-connect: communities are formed through plastic ingestion, cellular and physiological harm, as much as they are formed through external use, diffusion, and access. I would like to argue for plastic being responsible for a certain kind of posthumanism where our bodies are no longer materialized in the way they used to be before the "turn" arrived. We are plasticized, and our neomaterialist communities that are formed with the nonhuman are more secular than ever before. Plastic has developed an alterity wherein reducibility and irreducibility of the other is vexatiously problematized. I say, Fish becomes Man and Man becomes Fish. Momentously, the child born today, compared with the one born a century ago, is born *with* plastic and *into* plastic. Jacques talking about the Seven Ages of Man in Shakespeare's *As You Like It* would now speak about the Seven Ages of Plastic: from feeding bottles to toys to tiffin boxes to water bottles to writing desks to cars, to computers, to perhaps coffins.[61]

---

60. I have introduced this term into comparative literary studies. A fuller exemplification of "plastic literature" is provided in Chapter 4 of this book and also in my "Plastic Literature."

61. See Yarsley and Couzens, *Plastics*; also of interest is their "The Expanding Age of Plastics."

Plastic has come to build its own kinaesthetic habitat: it has the materiality of sensation and the autonomy of the other that calls for an ethical community. Denying plastic is not extirpating plastic, and living with plastic comes with the ethical quotient of reducing the harm and anxiety that it generates. Plastic has its intentions and demands confrontation through counterintentionality. It is exterior to us but it has reached our skin pores, punctured holes into our consciousness of health and healthy-mindedness, and become internalized within its own habitus. We are in the midst of a poetics of "plastic intensities."

Esther Leslie points out that "what is revealed . . . is the drive of the chemical industry towards 'the impersonation of life,' 'from death to death transfigured.' Refuse turns into worth in an act worthy of alchemy, but rather than cracking the code of life itself, all that has been achieved . . . is the polymerization of a few dead molecules. . . . Death imitates life and reinforces its domain."[62] How has the plastic affect brought us before a "philosophy of the undead" (the non-plasticity of plastic) and relentless disanthropogenization? Heather Davis perceptively notes, "Plastic suggests that we in the post-Kantian world have become voracious and solipsistic subjectivities driven by a dangerously self-interested will. Plastic, in this sense, represents the fundamental logic of finitude, carrying the horrifying implications of the inability to decompose, to enter back into systems of decay and regrowth. In our quest to escape death, we have created systems of real finitude that mean the extinguishment of many forms of life."[63] Plastic intensities have changed the way we think of "finitude." They have raised the question about how we can die, giving a fresh meaning to our philosophies of extinction. Somewhere death becomes deathless (call it plastic-death) and invites more and more forms of death-inducing moves—plastic thanatopoetics. In fact, plastic has stunned the anthropos, threatening to morph them within a circuit where human comes to surprise human. It works in a separate vein of freedom where plastic records our own choices to represent the world to ourselves and to see the world we knew and was familiar with get unworlded before us. This is another version of plastic unworlding. What kind of a historical change and paradigm shift are we facing in the Age of Plasticene? Plastic affect has brought us before a new love-poesis of plastic, telling us that it is

---

62. Leslie, *Synthetic World*, 25.
63. Davis, "Life and Death in the Anthropocene," 353.

humanity who creates plastic only to decreate humanity. It is a new creator at work, affecting on the definition of the human, humanist discourses, and the functioning of the planet system at large.[64] The affect works through what Dipesh Chakrabarty has called the new "figure of the universal,"[65] who is a creature in the becoming and grows out of an imminent catastrophe. This comes to challenge our self-identity. Plastic articulates the power of the invisible, the affect of the apparently discarded, and, consequently, the voiceless. The plastic affect repremises our zones of livability and responsibility; it can also be the inspiration to break free from the logic of plastic. Plastic is changing the history of our inheritance.

If plastic has simplicity, it is of the highest complexity—Nietzsche might have seen it as a marvel. This is elegant complicity, marvelous simplicity—a kind of plastic splatter where plastic builds ways of seeing beyond what we apparently get to see. Significantly, plastic is both artist and scientist. It came through as a solution to a problem, as a form to a certain set of requirements. It then offers its existence through a content, lives an idea, and sees its self-conjurement through registers of "mattering": first, plastic as plastic (the hermeneutic); second, plastic as something other than what we thought plastic is (the apophantic). It has already built an "involvement" beyond the formal limits of engagement. As a perverse variant of Ralph Cudworth's "plastic principle"—*nous hypercosmios* as against *nous enkosmios*—plastic now has created its own pervasive resonance that is complicit in material necessity, self-activism, auto-immunity, and an in-trending power. Art has not always been plastic; thinking always was. We try to stop thinking over plastic; but plastic "calls" out on us to think. We veered to plastic and all turnings became plastic, and, indeed, all re-turns too. The plastic turn, through its material-aesthetic plasticos, has unmade us and ungrounded the way we can think, express, and live.

---

64. Ronda, "Anthropogenic Poetics."
65. Chakrabarty, "The Climate of History," 222.

Chapter 2

Plastic Literary

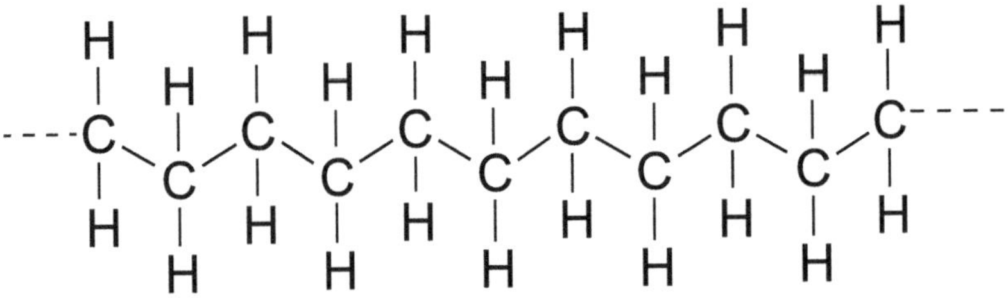

Figure 2. Polyethylene zigzag structure.

Polymeric structuration informs the core of twentieth-century critical consciousness. It is "plastic" in structure and property. With the modernist appropriation of cross-linkages, cross-overs, and intertextual paradigms such as adaptation, appropriation, pastiche, simulation, parody, and recto (recycling), we are thrown into the midst of plastic intertextuality and plastic avant-gardism. Twentieth-century thinking in arts and letters is deeply relational, flowing and oozing into and out of discourses that are apparently incongruous and incompatible. It is problematically invested in a plastic intertextual phenomenon. How does plastic become anthropophagic intertextuality and literary fractality, imbibing the principles of polymerism in twentieth-century thinking and writerly practices? Plastic in the critical consciousness of the twentieth century invests in imagination, desire, methodology, play, subjectivity, and malleability. It is polymerically aggregatory and combinatory, thinking through art and literature like a weave, bonding to form and transform in a variety of shapes and discourses.

Heterochain Textuality

Polymers and polymerization effectively produce long chains of molecules made up of a large series of interconnected links. It is the size of the molecules and their structures and physical states that determine the properties of the plastics, properties that determine their ability to get molded and shaped.[1] There are also the heterochain polymers in which compounds have oxygen, nitrogen, or sulfur atoms as their backbone chains. Thermoplastics, such as polyethylene and polystyrene, are molded and remolded repeatedly. It is the external heat that makes them change their shape and form. In thermoplastics individual molecules "are separate from one another and flow past one another"; they have the properties of "separability and consequent mobility."[2] Plastic responds to low temperature in developing complexity; not much hard effort is required, which means plastic can change its bonding and cross-over multiplication under low external changes (the external stimuli can be mechanical, thermal [enthalpy], and so on—mer [a unit] in bonding becomes poly). This demonstrates how plastic builds its own plasticity and is self-disintegrative. However, plastic nature is difficult to ascertain as it can be less elastic than rubber and can have varied properties of behavior when treated as either thermoplastics or cross-linked plastics.

In both versions the structural behavior and consequent changes and effect are different. Thermoplastics are linear chain polymers that soften on heating and solidify on cooling; cross-linked plastics are network structures (in three dimensions), which "once formed, are not softened by heating. The repetition of units in a polymer can result in a regularity of structure which may have important consequences; in favorable cases a number of chains may become aligned and in register for some distance, a state which is favored on energy grounds, and which is termed crystallinity."[3] Some properties depend on the "organic nature" that involves carbon bonding and combination (mostly considered "repeat units"), and some result from "the long-chain or network structure": "the former include density, dielectric permittivity,

---

1. See Rodriguez, "Plastic."
2. Rodriguez, "Plastic." See also Thompson et al., "Our Plastic Age"; Friedel, *Pioneer Plastic*.
3. Birley, Heath, and Scott, *Plastics Materials*, 1.

melting and softening temperatures, and the interaction with solvents, acids and bases. The polymeric nature leads to properties which are unique, exemplified by the very high viscosities of solutions and melts of linear polymers, and the accommodation of very high deformation without fracture, with the possibility of almost complete recovery on removing the stress."[4] It is, thus, *restless* with activity and possibilities.

Heterochain polymerization, within such behavioral and structural complexities, material-aesthetically, informs the dominant textual consciousness of the twentieth century. Plastic principle and behavior figure out the performance and pedagogy of major texts and theoretical-conceptual movements in reading. Literary-aesthetic thinking is embedded in a plastic molecularity that is developmental, mutative, mimetic, adaptive, and transgenic. Within the dynamics of the material-aesthetic hyphen, cross-linked polymerization may be seen as "cannibalism" as we find amorphous polymer-texts and art-work demonstrating and initiating unpredictable molecularity and viscoelasticity. Plastic polymers are "cannibals" because they alter and add (aggregation and agglomeration are interpreted as consumption) molecules for growth patterns that are remarkable and radical. The material-aesthetic dimension points out, for instance, a connection with anthropophagic poetics in that it ramifies and resonates in a variety of ways across the literary worlds, whether these be Latin American avant-gardism or European modernism.

Cannibalism, in this case, does not imply any "human consumption of human flesh." The word is used to refer to

a specific kind of ritual practice supposedly witnessed by European travelers to the New World beginning in the sixteenth century. Jean de Léry, along with others in his century and beyond, described a ritual cannibalism by which a group of people—such as the Tupinamba of coastal Brazil—would capture, kill, and eat a member of a rival group. The captors in that scenario would then claim to incorporate the power of the vanquished warrior, and the ritual would automatically invoke a reciprocal action on the part of that fallen warrior's group, should its members have the opportunity to capture one of their enemies in the future. That specific kind of ritual cannibalism took on partic-

---

4. Birley, Heath, and Scott, *Plastic Materials*, 2.

ular significance for Léry and other writers of his time, and it is in light of that sixteenth-century reading of that supposed practice that writers in later eras— especially Oswald de Andrade and Maryse Condé—use cannibalism as a metaphor for various kinds of cultural and textual incorporation.[5]

Such incorporations and appropriations are plasticogenic. The Brazilian poet-novelist and cultural critic Oswald de Andrade's *Cannibalist Manifesto* (1928) is deeply polymeric to the point of being amorphous; it is a "spaghettization" of literary and post-literary performances. Stephen Berg explains that "using the cannibal as a symbol, this manifesto recognizes the need for taking advantage of positive influences wherever they may come from and adapting them to Brazilian reality as a way of assimilating and organizing elements already saturated with cultural meaning. Imported cultural influences must be devoured, digested, and critically re-elaborated in terms of local conditions."[6] Cannibalism also speaks of a tropicalist mindset where eating the other is not necessarily about being eaten by the other; maintaining identity without being swept away by the other, the anthropophagic perspective advocates nationalist identity as an "open access" and an open-ended process. This is not about an unabated quest for origins but a contested zone that cross-links cultural-aesthetic influences from areas and sources other than one's own.

It is an intense plastic desire—plastic's ability to elaborate and re-elaborate itself—that, through Brazilian Concrete and Neo-Concrete art and Tropicália, brought Brazilianism and the non-Brazil into "high art": works with high appropriative power that enabled intersections of thoughts, figures and trends in music, art and visual culture. In contrast to poet-critic T. S. Eliot's plastic art (as explained later in this chapter through his *The Waste Land*) or German artist Kurt Schwitters's *Merz*, where there is an outward polymerization in the sense of "reaching out" and branching out, *Manifesto Antropófago* declared branching in (crystalline spherulites, as it were) that collapsed Shakespeare, Montaigne, Freud, Picasso, constructivism, Gestalt psychology, and other issues into its impulsive mastication—at once endo- and exo-cannibalism brought into a fierce force-field of assimilation. Both kinds of writing demonstrate covalent

---

5. O'Reilly Tonks, "Cannibal Routes," 7.

6. Berg, "An Introduction to Oswald de Andrade's *Cannibalist Manifesto*," 91.

bonding;[7] combined, they speak about a molecularity that has kinetic and potential energies. Thoughts that come together demonstrate "ground-state energies" where even a local thought-particle—literary-aesthetically we may qualify it as a "fragment"—cannot be left in sedation and isolation.[8] It may be argued that covalent bonding in such modernist consciousness brings thought-atoms together, keeping a stable molecule (a dominant thought or idea) in place but not without its dissociation energy.

If we were to rewrite de Andrade's manifesto in terms of plastic,[9] it would probably read like this:

I am only concerned in what is not mine.
Law of Man. Law of the cannibal.
I am concerned only in what is not human.
Law of the material, the more than human. Law of the plastic.

What clashed with the truth was clothing, that raincoat placed between the inner and outer worlds. The reaction against the dressed man.
What clashed with the truth was plastic, that impermeable coating that stood between the inner and outer worlds.
The reaction against the plastic man.

It was because we never had grammars, nor collections of old plants. And we never knew what urban, suburban, frontier, and continental were. Lazy in the *mapamundi* of Brazil.
It was because we always had grammars, and collections of old plants and freshly experimented ones. And we always knew what was urban, suburban, frontier, and continental. Restiveness in the mapamundi of Synthetica.

A participant consciousness, a religious rhythmics.
A polymeric consciousness, an integrative rhythmics.

---

7. "Covalent bonds are the most important means of bonding in organic chemistry. The formation of a covalent bond is the result of atoms sharing some electrons. The bond is created by the overlapping of two atomic orbitals" (Stauffer, Dolan, and Newman, *Fire Debris Analysis*, 54).

8. Bacskay, Reimers, and Nordholm, "The Mechanism of Covalent Bonding," 1494.

9. De Andrade, "The Cannibalist Manifesto."

Routes. Routes. Routes. Routes. Routes. Routes. Routes.
Mers, Mers, Mers, Mers, Mers, Mers, Mers.

I asked a man what the Law was. He answered that
it was the guarantee of the exercise of possibility.
That man was named Galli Mathias. I ate him.
I asked a man what the Law was. He answered that it
was the potential for the exercise of possibility.
That man was named Plastic. He ate us.

From William James and Voronoff. The transfiguration
of the Taboo into a totem. Cannibalism.
From William James and Plastic Sea.
The transfiguration of the Taboo into a totem. Plasticism.

We are concretists. Ideas take charge, react, and burn
people in public squares. Let's get rid of ideas and
other paralyses. By means of routes. Believe in signs;
believe in sextants and instars.
Plastics are plasticists. They make ideas and images that take over.
Let us allow plastic to change our itineraries.

Under plastic-cannibalism, Oswald de Andrade and the modernist artist Tarsila do Amaral intertextualized European modernism and growing trends of avant-gardism with indigenous trends in thought and art. Tarsila, as Stacey Wujcik explains, "began exploring themes of mulattos and blacks, Brazilian folk culture, and the urban *favelas* (slums). Many of her subjects were drawn from the stories she heard from the 'black slaves on her grandfather's coffee plantations.' Though her subjects were Brazilian and painted with the 'intense colors' then associated with Brazilian art, Tarsila worked in an individualized style of Analytic Cubism she observed in Europe."[10] Itineraries such as plastic malleability began changing. For instance, her painting of a nude figure which de Andrade titled *Abaporú* (meaning "man eats," derived from the dictionary of the Tupí-Guaraní) polymerizes Cubist abstraction, Brancusi's art, with Brazilian stylistic indigeneity. The cannibal

10. Wujcik, "Oswald de Andrade's Manifesto Antropofago in Brazilian Anti-Art and the Works of Cildo Meireles."

became a recurrent operative metaphor and token of plastic desire even among the Dadaists and surrealists as the co-constructive strategies increasingly held sway among artists across continents.

On such co-habitative and co-occurent notes, conceptual Brazilian artist Cildo Meireles's art can be considered as multisensorial investments that open up the space "in the physical, geometric, historical, psychological, topological, and anthropological sense."[11] Working on anthropophagic reasoning and energy, the space is rendered prismatic: it works on different patterns of "coming together" as instanced through Meireles's *Southern Cross, Missions, Glove Trotter*, and *Babel*. For instance, the sculptural installation *Babel* (2001) intertextualizes the biblical story in Genesis with "a five-meter-high tower formed by some 700 radios from different times, ranging from large, valve radios to transistors, all tuned to different stations"[12]—material-aesthetically the plastic chains, the branching and tuning across. The installation involves hundreds of "second hand analogue radios [that are] stacked in layers."

Meireles calls this "the tower of incomprehension" as the radios connected to a variety of stations are all played together in a minimum volume resulting in a continuous, though muffled, sound, indistinct and somewhat discordant—a "noise" that has its own meaning-sounds. The noise is "plastic." The polymer-chain principle informs all the radios "of different dates, the lowest layers nearest the floor being composed of older radios, larger in scale and closer in kind to pieces of furniture, while the upper layers are assembled from more recent, mass-produced and smaller radios."[13] Here Meireles polymerizes Borges's "The Library of Babel" and "The Aleph" with references to the Mesopotamian tower of Babel and issues of memory whereby people can connect with their radios from the past. Moacir dos Anjos points out that the presentation of informational over-layering in *Babel* can be seen as a metaphor "for the intricate relations between distinct nations and communities," one that insists on "recognizing the existence of a territory with uncertain boundaries, one that accommodates multiple oppositions and

---

11. Meireles, "Artist's Writings," 136.
12. "Cildo Meireles."
13. "Cildo Meireles, Babel, 2001," https://www.tate.org.uk/art/artworks/meireles-babel-t14041, Tanya Barson, May 2011.

Figure 3. *Babel* (2001). Photo by Sophie Mutterrer. Used by permission of the artist.

produces the multiple contamination of cultural expressions previously separated by geographical and historical injunctions."[14] This is a new-found alterity conceived, as Stephen Home argues, through a "constellation of practices in which individual bodily well-being and the social/political are experientially defined in relations of reciprocal exchange with each other."[15] The material-aesthetic intertwining brings artistic self-reflexivity in a conglomerate of sensuality, experientiality, embodiment, and poetic language—the "political" and "poetic" caught in a warm embrace.

Hetero-chain textuality makes for a profound presence when one sees how a "doodle" that Meireles drew in 1976 resulted in *Malhas da Liberdade* (*Meshes of Freedom*). *Malhas da Liberdade*, which Meireles considers his favorite artwork (Figure 4), comes as a "waterfall of bifurcations because it somewhat shows how my mind operates: always shifting from one thing to another, bifurcating."[16] Meireles explains that the "first version began as a doodle. I happened to see a fisherman weaving his fishnet and asked him to do a net following the procedure I had drafted. When I did the show at London's Tate Modern I decided to do the fourth version of the work in plastic units so people of all ages could help build the meshes. The idea is that people could join to build a large version of this interactive open structure."[17] Encouraging "interactive concepts of mode and medium" to bring about a well-networked articulation in "time, resistance, situation, memory, measure, space, language, emotions especially fear, and, of course, politics," he ends up producing a densely layered polysensorial space.[18]

The doodle entangles itself generously into a "grid," a sort of reticulation in metal, thread, and plastic, in participation and imagination, to create a "system of bifurcation or divisions and duplications." The plastic-literary of *Meshes of Freedom* cannibalizes mathematics (as in Mitchell Feigenbaum's work on chaos theory), fractality, geometry, physics, pattern-studies, ideas of participatory democracy, spectrality, narratives of dictatorship and emancipation, processes of communication, history, and other disciplines, bringing it closer to the spaghetti-like structure of the polymer chain.

---

14. Anjos, "Where All Places Are," 173.
15. Home, "Cildo Meireles," 43.
16. Meireles, interviewed in Garcia, "The Scale of Common Things."
17. Meireles in Garcia, "The Scale of Common Things."
18. Garcia, "The Scale of Common Things."

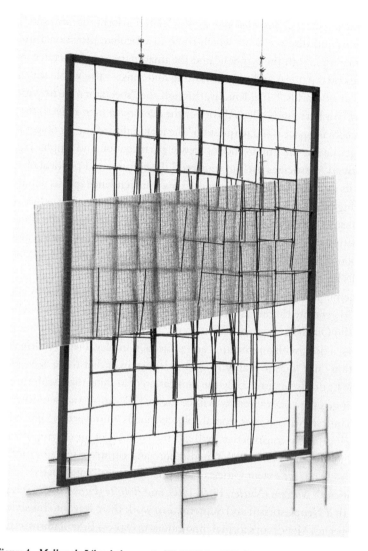

Figure 4. *Malhas da Liberdade*—versão III (1977) by Cildo Meireles. Photo by Edouard Fraipont. Used by permission of the artist.

Both consist of high "repeating units" that produce the entanglement, intersections, "mesh pattern," middle-growth, and limits to form an operative poetics. Also, both forms surface a fear of the labyrinth: the known and the unknown, the foreseen and the unforeseen, in the "branching off." Meireles's meshes are plastic covalencies as we encounter the opening of passages, "interlacing places," suspension of a completely a priori program, and "combinations and displacements which come to articulate provisional units."[19] The material, both the symbolic and the imaginary, conglomerates to produce its own pattern and poise, meaning and sense—the whole act caught in a war of worlds.[20] Incidentally, through the "spectator's participation in the art" (in Ferreira Gullar's theory of the *não-objeto* [non-object]) the neo-constructivist space starts to question "the materiality of the art object"; such linkages lie out the importance of viewer-participation and emphasize how an artwork plasticizes itself to transcend the cultural and political objectivity of its establishment and existence. This transcendence speaks about a re-ordering of the notion of time and space and also place and structure. Plastic transcontinental desire re-elaborates literary planetarity, and the vibrations and activations across continents form a new poetics of absorption and adsorption. This is a sort of "juggle" among a variety of options that Meireles installs his faith in—the artist for him is always a "juggler."[21]

Chemo-aesthetically speaking, the principle of breaking the chain of polymers to generate fluidity is analogous to the "constructive" in Brazilian artist Hélio Oiticica's trajectory. This constructive is not constructivism; it is a process, a flow, with an intention and projection in play. Piet Mondrian's observation on a "plastic reality"—"What is certain is that there is no escape for the figurative artist. . . . By the unification of architecture, sculpture and painting, a new plastic reality will be created. Painting and sculpture will not manifest themselves as separate objects, nor as 'mural art' or 'applied art,' but being purely constructive, will aid the creation of an environment not merely utilitarian or rational, but also pure and complete in its beauty"[22]— explains to a large extent Oiticica's formulation of the "constructive."

Oiticica's *Núcleos* (*Nuclei*, 1960–1963) and *Bólides* (*Fireballs*, 1963) is plastic in that Neoplasticism and Suprematism work their way into his style. He appropriates Mondrian's formal innovations to express Brazilianism as much

---

19. Scovino, "Tactics, Positions and Inventions," 439. Meireles states, "The journey through a labyrinth is premised on a thoughtful, attentive search. You have to walk, but with each step, you have to stop and think. That is the good thing about labyrinths. They help us to slow down" (in Farmer, "Through the Labyrinth,"), 41.

20. Appropriated from Orson Welles's radio broadcast, *The War of the Worlds*.

21. Herkenhoff, *Cildo Meireles*, 21.

22. Piet Mondrian, "Plastic Art and Pure Plastic Art," quoted in Martins, "Hélio Oiticica," 410.

as to speak of issues concerning Brazilian art and times. There is a kind of rhythm across traditions, a rhythmic movement in color, emotion, space, line, and visuality. It is in the "constructive destruction" that the flexibility and liquidity of plastic chain entanglements lie. The "constructive" deeply bears out the plastic reality as Oiticica points out: constructive artists are "builders of structure, of color, of space and of time"; they add "new perspectives" that modify the ways in which "we see and feel." The "constructors" open up "new directions in contemporary sensibility."[23]

This constructive aspect of plastic reality shows up in European modernism through the work of several artists, one of whom is Kurt Schwitters. Oiticica's *Parangolé* and Schwitters's *Merz* are profound polymer-projects as both the artists experiment in plastic polymericity through a variety of objects and "appropriation of environments and places." Kurt Schwitters believes that art exists only in relationship to others ("reciprocal illumination of arts")—a variation of plastic anthropophagy. In his *Merz*, like the Cubists and the Futurists, he brings experimentation through what others characterize as refuse or rubbish, minimizing the importance of the extra-aesthetic content (*Eigengift*) of his materials and bringing about their artistic transformation (*entmaterialisiert*). Schwitters's plasticity lies in what he means by "artistic forming" (following up on German romantic theories of intentional fragment): sustained investments in abstract art, typography, graphic design, and collage arts. Aesthetic and self-referential plasticities put Schwitters in polymeric conversation with the Bauhaus professors Josef Albers, Herbert Bayer, and László Moholy-Nagy; members of the avant-garde groups and movements De Stijl and Berlin Club Dada, such as Theo van Doesburg, John Heartfield, and Hannah Höch; and the proponents of New Typography, such as Jan Tschichold, Paul Renner, and Max Burchartz—*a polymeric consciousness, an integrative rhythm.*

Schwitters's *Merz* is, in fact, overwhelmingly plastic: even the name comes about through an intertextuality that sees Schwitters chancing upon a name Kommerz-und Privatbank, homonymic with *kommerz* (commerce) and also *Herz* (heart), *Scherzo* (a prank), or *Marz* (March).[24] It is in the ensemble and the folds that Schwitters finds his plastic Dadaism, as he technically and

---

23. Oiticica, "The Transition of Color from the Painting into Space and the Meaning of Construction," 224.

24. Kłos, "MERZ."

aesthetically brings conceivable materials together—playing card, tram ticket, envelope, shoe lace, newspaper, wallpapers, prints, labels, string, brush or pencil stroke, wire net, painting, and so on—to evolve an expression that can elicit annoyance, reproach, wonder, admiration, and confusion. *Merzbild* paintings are plastic collages, polymerizing a variety of materials in their recurrence, co-existence, and perceptive proximity. They work at the limits of expression and systems of thought. Materials textualize each other into forming a world that announces a new aesthetic: the plastic-dadaistic. *Merzbau*, intergenerically, came with its own cannibalism as the construction demonstrated

> numerous niches into which Schwitters systematically put different, redundant items: a lock of hair, an unimportant scrap of paper or a vial with urine. These things came for the artist's family and friends. There were niches of Mondrian, Arp, Doesburg, Malevitch and many others. When the niches were completely filled, Schwitters's bricked them up and created new ones. And so *Merzbau* grew. And since this *opus vitae* was being built in Schwitter's own, three storey tenement, soon the ceiling became an obstacle to the further upward growth of the construction. Not thinking much, the artist simply pierced the ceiling and continued, later breaking through the other two ceilings as well. *Merzbau* was a kind of collage, assemblage, it was both *Merz* and *Merzbild*, a shift or removal of the border between art and life.[25]

Merz is one plastic moment that intensified and insinuated everything: *Merz* was the cannibal and when Schwitters wrote in 1927, "Now I can call myself *Merz*," the creator became the cannibal.

If Schwitters can bring a panorama of objects into the constructive, Oiticica, on the other continent, brings shacks used in construction sites, "popular cubbyholes and constructions, generally improvised, which we see every day, also, fairs, beggars' homes, popular decorations of traditional, religious and carnival feasts, and so on."[26] The "constructive" evolving as the amorphous structure of a polymer. Through the tangle and complexity of the chain reactions that make life and art intersect at critical and delicate angles, Oiticica seeks to "emphasize the fluidity of life in opposition to fixing and sys-

---

25. Kłos, "MERZ."

26. Hélio Oiticica, "Notes on the *Parangolé*," in *Hélio Oiticica* (Paris: Galerie Nationale Jeu de Paume, 1992), 87–88.

tematizing the world."[27] This plastic assemblage works out also in Oiticica's *Tropicália* (1967), which talks about a

> total environment, or *ambiente*, consisting of two simple structures with cloth siding called *penetráveis* ("penetrables") surrounded by tropical plants, dirt paths, and live parrots displayed in cages. The two *penetráveis* evoked what he called the "organic architecture" of the favelas, although they were not designed as reproductions of shanties, which were typically made of wooden planks or even bricks and mortar. The larger penetrable was a spiraling structure that led the participant through a dark passage to a functioning television set, thereby registering the presence of the then-new technologies of mass communication in the poor urban communities.[28]

The inter-material pattern demonstrates "hysteresis," which in polymer sciences refers to changes that come because of previous events, through co-presence and collateral impact, "leading to results that don't seem to be reproducible in the way that we would normally expect."[29] This is plastic anthropophagy that produces suspension, extension, rupture, and transgression.

Oiticica's *Parangolés* is a similar case in addition, condensation, in-side chain branching, and cross-linked configuration. In fact, as Renato Rodrigues da Silva notes, "Oiticica not only incorporated objects and procedures into the *Parangolé*, but also the distinct universes of avant-garde experiments, samba, and his personal life. The artist's performances of the proposition in the Favela da Mangueira, the Museu de Arte Moderna of Rio de Janeiro, and the Aterro do Flamengo Park, among other different contexts, direct us to investigate a doubtful unity."[30] *Parangolé* point towards a conceptual unity but ensures a restlessness; this makes possible the plastic phenomenon of "branching out" that encourages *couple des contraires*—"interior and exterior, beggar's rags and king's robes, water bag and bed linen, masculine and feminine."[31] It is a trans-motion among paradigms of sexuality and culture, the multisexo, two-wayness (bi-directionality)—expressivity as *la volonté de*

---

27. Epps, "Hélio Oiticica."
28. Dunn, *Contracultura*, 79.
29. Schaller, "Polymer Properties."
30. Da Silva, 'Hélio Oiticica's *Parangolé* or the Art of Transgression," 214.
31. Da Silva, "Hélio Oiticica's *Parangolé* or the Art of Transgression," 217.

susciter l'expressivité.[32] Plastic cannibalism cannot allow "Eruditamos tudo" ("Eurocentric erudition," in Oswald de Andrade's sarcastic description of Brazil's lettered class),[33] as it declares a codification, a bricolage that must relate creatively with the Brazilian world.

In this context, Oiticica's *The Eden Plan* (1969)—which the artist described as "an exercise for the creleisure and circulations"[34]—brings a polymeric compatibility with creative activity and everyday life. "Creleisure" is about creating *through* leisure, as crefeel, cresleep and credream—all of these point to an integration of a variety of narratives, an experimentality in "convi-convivência" (a neologism Oiticica coined in his 1970 essay, "Brasil Diarréia," to describe a hypocritically complicit coexistence with the nation's inequalities), which demystified the known and the tested. On an ambiguous note, Oiticica points out that creleisure can be a creation of leisure or belief in leisure ("I don't know, maybe both, maybe neither").[35] The concept of creleisure has a revolutionary character to it, and, as Christopher Dunn notes, "the tropicalists would give impetus to emerging countercultural attitudes, styles, and discourses concerning race, gender, sexuality, and personal freedom."[36] Both *Parangolé* and *Merz*, thus, build themselves on a totality—moving between amorphous and semi-crystalline phenomenon—which comes from an experience of the unfolding elements that constitute the art-work, the participatory quotient and art-situation. They dwell on a plasticization that explores the paradox of formulation and "non-formulation of concepts," keeping the transformative process on both ends (art and the spectator) active. These plastic texts are predominantly "evental, intersemiotic assemblage that brings into play a multitude of non-poetic regimes of signs."[37] The Brazilian critic Teixeira Coelho sees a tension in everything while "confrontations, eventual 'swallowings' or incorporations ('recuperations') always leave marks in both the elements."[38] Juxtaposition or agglutination

---

32. Da Silva, "Hélio Oiticica's *Parangolé* or the Art of Transgression," 217.

33. Shellhorse, "Subversions of the Sensible," 165.

34. http://amper.ped.muni.cz/~jonas/knihy/10_instalace_jako_strategie/Claire%20 Bishop,%20Installation%20Art.pdf, 108.

35. Skrebowski, "Revolution in the Aesthetic Revolution," 66.

36. Dunn, *Brutality Garden*, 3.

37. Shellhorse, "Subversions of the Sensible," 167.

38. Araujo, "Postmodernist Intertextuality," 5.

marks the plastic anthropophagy, keeping invention and intervention—Denise Araujo prefers to call this "aesthetics of hypervention"—on a delicate axis of operation. Whether it is the "experimental exercise of freedom"[39] (coined by Mário Pedrosa) or the "aesthetics of hunger"[40] (in the words of Glauber Rocha), or Oiticica's idea of "Brasil Diarréia," we are in the midst of plastic performativity and equilibrium, which are deeply marked by countertextuality.

Crystalline-Amorphous Textuality

On that note of permeability and reorganization, interpenetration and cross-linked systems, can we consider Eliot's *The Waste Land* as polyethylenic?

The "inter" properties of plastic express through chemical composition, the temperature-pressure variability, the fluctuation-gradients, the additive factorizations, and other factors. Plastic provides a schema and web of communication: it provides through its multiple polymeric cross-overs and organic doubling the complex equations of blending and mixing with the different, the transformed and the conflictual. *The Waste Land* in all its five sections becomes a "cross-over" figure *quoting* across literature and discourses in a kind of circularity of reading, as Roland Barthes has argued. Patterns also exist as the amorphousness of links and linkages. There is an "anxiety of connection" as the poem changes levels of understanding and produces different polymers out of its growth. In its semi-crystalline status, each section of the poem executes a covalent bonding to create co-polymers. The co-polymericity manifests in how Dante, Shakespeare, Baudelaire, the Bible, and Webster intersect with Siegfield Follies, bawdy camp songs, tarot cards, and contemporary newspaper accounts. Jayme Stayer notes the "large number of references to the great and small events, the fads and fashions of the epoch such as the contemporary cult of Wagner, occultism, the grammophone, rag time music, cigarettes, motoring, the new sexual morality, pollution, the preservation of old churches, demobilization, the effects of the

---

39. https://www.moma.org/momaorg/shared/pdfs/docs/publication_pdf/3232/Pedrosa_PREVIEW.pdf?1456334850, 229.

40. https://www.amherst.edu/media/view/38122/original/ROCHA_Aesth_Hunger.pdf.

First World War, and the decline of the West generally."[41] Richard Baden-hausen points out that

> Eliot's criticism is a vocabulary that incessantly evokes, suggests, or depends on collaborative activity, for that lexicon is distinguished by repeated uses of words like "amalgamate," "assimilate," "balance," "collective," "collocation," "combination," "composite," "compound," "converge," "cooperate," "doubleness," "harmonize," "mixture," "reciprocal," "reconciliation," "syncretism," "synthesis," and "union." Eliot's favorite expression marking this process is "fusion," which surfaces remarkably in over thirty separate essays. This suggests Eliot was particularly preoccupied with the notion of exploring the productive unification of distinct elements and he tended to view the creative process in terms that located some type of collaboration as its fundamental, guiding principle. Whether he is discussing the make-up of the poet's mind in terms of a chemical analogy in which two gases mix in a productive combination or examining the rejuvenating union of a translator's mind with that of the original artist, Eliot assumes that distinct entities must come together for creation to occur.[42]

The poem-polymer is, in fact, experimental and heterophonic.

On that note of doubleness, compound, composite, composition, and synthesis we can take the instance of high-density polyethylene (HDPE), which is composed of macromolecules where the long carbon chains are our points of attention. The behavioral manifestation of high-density polyethylene and low-density polyethylene is due to macromolecular structure. HDPE is a rigid translucent solid "which softens on heating above 100° C, and can be fashioned into various forms including films. It is not as easily stretched and deformed as is LDPE."[43] LDPE, on the other hand, is a "soft translucent solid which deforms badly above 75° C. Films made from LDPE stretch easily and are commonly used for wrapping. LDPE is insoluble in water, but softens and swells on exposure to hydrocarbon solvents. Both LDPE and HDPE become brittle at very low temperatures (below −80° C)."[44]

---

41. Stayer, "The Dialogics of Modernism," 98.
42. Badenhausen, *T. S. Eliot and the Art of Collaboration*, 9.
43. "Properties of Polymers."
44. "Properties of Polymers."

The morphological changes contribute to these variations as polymer molecules are large and get packed together in a nonuniform fashion "with ordered or crystalline-like regions mixed with disordered or amorphous domains."[45] There is a tangled existence, and the degree of crystallinity is influenced by three factors: chain length, chain branching, and interchain bonding, all of which significantly constitute the plastic intertextuality (in this context, the text-polymers). *The Waste Land* morphologizes on the macromolecularity of concepts, ideas, and contexts through a chain length, chain extension, branching, and bonding across other texts as they come together in alternate domains of crystallinity (branched chains) and amorphousness (long unbranched connections). Plastic texts such as *The Waste Land* are a mix of high crystallinity and amorphous regions, demonstrating differing levels of permeability and polymer repeat units (as part of the polymer, they continue to bond, which leads to the formation of the complete polymer). Textual polymerization is stiff (definitive) and flexible (accommodative) at the same time, behaving both as LDPE and HDPE. The poem, like its counterparts in Joyce's *The Portrait of an Artist as a Young Man* or *Finnegan's Wake*, conducts a "chain transfer" that diverts chain growth as we encounter inter- and intramolecular thought-formations in a variety of ways and routes: *mers, mers, mers*.

The intramolecular pattern here demonstrates radical site-generation as side-chain radicals lead to the formation of additional side chains through chain transfer reactions. This is the amorphous space differential, the generative morphology coming out of polymer branching. The "Burial of the Dead," "The Game of Chess," "The Fire Sermon," "Death by Water," and "What the Thunder Said" are at once unique in themselves and not. All demonstrate "high intensity macromolecularity" as each of the five polymer-units (say, ethylenic units) generates intramolecular meaning and releases it to the other as a side-chain radical formation. This process produces the semi-crystalline poem with its chain entanglements and "differentials," which we know as *The Waste Land* (the polyethylenic). It is through these five polymeric units that *The Waste Land* as radical polymerization begins generating nearly unchecked chain-transfer reactions. For instance, "Death by Water" initiates a chain transfer where a thought (technically could be an intramolecular hydrogen atom transfer) that could be symbolic-metaphoric in nature, say, "water and sterility,"

---

45. "Properties of Polymers." 2019.

transfers to another polymeric unit in "What the Thunder Said" ("rock and water"), producing a radical chain growth in poetic macromolecularity. The co-polymeric growth leading to radical chain reactions comes through "water" as thought-molecule transfer. Every section is long chained and is deeply polymerized in "meaning units" and "concept bonds" as they connect together to form a supramolecular chain status that can often be coiled or twisted. Each has "meaning stress" to it. It is interesting to note in this context that "when the polymer flows, the polymer chains begin to slip, or slide, over each other while rotating around the carbon-to-carbon bonds. Layers slip one over the other and this is called shear or laminar flow. As the individual polymer molecules move relative to each other, they may change their orientation or direction due to friction between layers and chain entanglements."[46] "Death by Water" slides and slips and flows into "What the Thunder Said," ensuring "states of meaning" that are crystalline and amorphous; often intertwined, they can also exist in a random fashion and fold upon each other.

The tangling and twisting can be further localized through a "chain packing density," for instance, in the character of Tiresias as he figuralizes his own way into the macromolecularity of the poem. Such instances go to show how ideas and conceptual formations can have a "low chain persistence,"[47] which exposes ideas to flexibility and polymeric conjugation. Tiresias becomes a "plastic composite" having its own status in the poem and in the texts of Sophocles, Ovid, Homer, and Dante. The chain density grows through borrowings from Apollinaire's *Les mamelles de Tiresias*, Swinburne's "Tiresias," and Milton's *Samson Agonistes*. Katherine Spurlock explains that "in the controversial note to Tiresias, Eliot displays a desire to abdicate the author's conventional role as 'unifying' consciousness, turning over that responsibility to his composite, reticent prophet who, although 'not indeed a "character," ... sees ... the substance of the poem' and author-like 'unit[es]' 'all the rest.'"[48] Spurlock also notes that Eliot "wished readers to associate the consciousness 'behind' *The Waste Land*'s string of psychically distressed speakers with an anti-Oedipal, Tiresian observer. Thus, from a psychoanalytic standpoint, Tiresias does not only impersonate a major character because,

---

46. Dynisco. "Polymers and Plastics."
47. The persistence length is the property that quantifies the stiffness of a polymer; low chain persistence signals flexibility.
48. Spurlock, "Rival Authorities," 236.

as a composite made up of so many texts, he resists being read as a projection of any particular psyche."[49] This speaks of chain overlap. In fact, the chain persistence of the poem is not rigid, and dense interactions happen at various levels—whether in such "local turbulence" as in Tiresias or tangled transitivities as in "The Burial of the Dead"—in the formation of macro-molecularity. Plastic intertextuality becomes a complicated process whose internal mobilities are often difficult to determine and denote. The poem builds its own "equilibrated configurational freedom" but not without moderate chain persistence. Cross-links keep the poem "busy" as we encounter the principle of transposition too. Plastic intertextuality opens up the text to its internal formations and potential dimensions of meaning; it is a commentary on the politics and poetics of textual composition as well.

Not that such intertextualizations demonstrate the workings to the full. Textual plasticities can leave certain formations not fully explained. These are the drive and inherent instability of plastic intertextuality. *The Waste Land* plasticizes what Barthes calls "the stereographic plurality of its weave of signifiers"; it is stereophonic in citations, echoes, references, and appropriations, which further reminds us of his insight that "on the one hand the stability and permanence of inscription," is "designed to correct the fragility and imprecision of the memory, and on the other hand the legality of the letter, that incontrovertible and indelible trace, supposedly, of the meaning which the author has intentionally placed in his work; the text is a weapon against time, oblivion and the trickery of speech, which is so easily taken back, altered, denied."[50]As a semi-crystalline event, the poem is close to being a "text of bliss," relating issues that are historical, cultural, philosophical, and aesthetic and connected to taste and values. Plastic polymerization is structuralist to a large extent where there is an emphasis on systematic and relational aspects of "coming together"—bricoleur in logic. Barthes's way of seeing plastic has a structure: bricoleuring the orientation through metaphors, motifs, references, images, and other linguistic signs. The text here does not work as a unitary whole, for articulations beyond the immediate structuring of the text

---

49. Spurlock, "Rival Authorities," 237. In contrast to the way I see Tiresias as a plastic composite, Matthew Scully finds Tiresias as a plastic figure and frames his connection, through Catherine Malabou's "plastic reading," with the plasticity of *The Waste Land*. See Scully, "Plasticity at the Violet Hour."

50. Barthes, "Theory of the Text," 32.

refuse to allow the system of signs to stay enclosed. Plastic production is structural but slightly autonomous and somewhat unpredictable as well—the crystallinity-amorphous phenomenon.

Plastic intertextuality cannot ignore Gérard Genette's "transtextuality" in that certain types of discourses and formations become "open structuralism" themselves. It is a structure that builds a discourse and classification but not without branching out bonds that extend the circle of formality of structuration. *The Waste Land* typifies such a formation. Plastic intertextuality maps and is not closed to revising its own lines of mapping; it makes us rethink the poetics of expression. The transtextuality of *The Waste Land*, like many other major texts of European and Latin American modernism, declares patterns of relations and enunciations—a complicated poetics of being-with. It makes us rethink the dynamics of co-presence and a totality of relevant relationships. This whole is not total confinement but a semi-autonomous field that has a desire hidden inside its systemic layers. This desire is crystalline-amorphous. Just as plastic keeps constructing its own plasticity, *The Waste Land* builds its own trace as well. The plastic in the poem is the trace; and intertextuality oversees how it "explodes" and "disperses" (in the words of Barthes). Both as transtextuality and metatextuality, where the relation is in a text uniting a "given text to another, of which it speaks without necessarily citing it (without summoning it), in fact sometimes even without naming it,"[51] the poem comments on other texts and builds a unique relationship with them. In so doing, it confirms a relation where every bond or network connects with the preexistent, builds on and yet does not overlook the former, leading to multiple bonding, chain setups, and chain transpositions.

Plastic is under trial as external changes and internal movements keep it in motion. Textual consciousness through the twentieth century is under trial too. The plastic-text keeps up its own trial-as-formation across structural varieties and semiotic practices. Julia Kristeva reminds us about seeing a text as rearranging itself, its inside with its outside, and what it brings with a preexistent system of meanings. For Kristeva—as for Mikhail Bakhtin, too—it is the dynamic of textual growth and meaning-proliferation in both horizontal (linear polymerization) and vertical dimensions (non-linear branching out). Kristeva writes:

---

51. Genette, *Palimpsests*, 4.

Dialogue and ambivalence lead me to conclude that, within the interior space of the text as well as within the space of texts, poetic language is a "double." Saussure's poetic paragram ("Anagrams") extends from zero to two: the unit "one" (definition, "truth") does not exist in this field. Consequently, the notions of definition, determination, the sign "=" and the very concept of the sign, which presupposes a vertical (hierarchical) division between signifier and signified, cannot be applied to poetic language—by definition an infinity of pairings and combinations.[52]

Plastic growth never leaves a text unitary. *The Waste Land* or *Waiting for Godot* or *Ulysses* can never be one. The Kristevan plastic sees a systemic discursivity interfused with ambivalence; the plastic behavior is not always strictly logical and rigorously predictable. The plastic poetic, for Kristeva, works on doubling, the cross-over listing, where both "A" and "notA" form a structure, a dialogic bonding that is both chemo-poetical and plasticogenic. Kristeva sees in such a relation a poetic double where opposition between the monologic or carbon-hydrogen unit is one of tension and tensility—an identity with disparity. There is a prohibition to collapse into each other and a tension to stay formally polymeric. The plastic-text is a text in conflict, in opposition, analogy, coherence, and logic—both as phenotext and genotext.[53] It is not a strongly totalitarian entity. The essential plastic-nature builds the contradiction and the promise of dialectical transcendence, among competing atoms of

---

52. Kristeva, *Desire in Language*, 69.

53. Referring to Kristeva's "L'engendrement de la formule" (in *Semeiotikè: Recherches pour une sémanalyse* [Paris: Seuil, 1969]), Johanne Prud'homme and Lyne Légaré explain these two terms as follows: "From the Greek *phainesthai*, 'phenol' means 'to appear.' The term 'pheno-text' refers to the text as a 'fact' or an 'appearing' in its concrete manifestation or material form (communicative function). It is the site where a space for the process of engendering meaning is embodied in a concrete medium. It acts as the focalizing point of the signifying process. The printed text is where the production of meaning is momentarily suspended" ("Semanalysis: Engendering the Formula," n.p.). This is one part of plastic structurality. They continue, noting that the geno-text comes from "the Greek *genêtikos*, 'geno' represents that which is 'specific to generation,' in the sense of 'genesis' and 'production.' The geno-text corresponds to the process of generating the signifying system (the production of signification). It is the locus of all possible signifiers in which the formulated signifier of the pheno-text (the formula) can be situated, and thus, overdetermined. All of the possibilities of language (the symbolic process, the ideological corpus, the language categories) are arrayed there before precipitating out in some formula in the pheno-text. The geno-text is not a structure; it represents signifying infiniteness. The geno-text does not reveal a signifying process; it offers all possible signifying processes [*signifiances*]." This covers for the two major facets of plastic thinking and plastic behavior where one concretizes (the matter) and the other finds itself in a process of emergence (mattering). Both are *signifying* in their own distinctive identities.

thinking, molecules of thought; this is the unity in disparity. Plastic entanglement becomes an event shot through with (in Bakhtin's formulation):

> shared thoughts, points of view, alien value judgements and accents. The word, directed toward its object, enters a dialogically agitated and tension-filled environment of alien words, value-judgements and accents, and weaves in and out of complex interrelationships, merges with some, recoils from others, intersects with a third group: and all this may crucially shape discourse, may leave a trace in all its semantic layers, may complicate its expression and influence its entire stylistic profile.[54]

Linkages matter; and changes brought into the polymeric structure through heat and density transform the character and expression of plastic. As with language, so with plastic: we encounter stratifications and seemingly endless interpretations (polymeric expressions) that are built through a series of moments, shifting positions, conflicts (class, social and molecular), and discourse (speech-acts, carbon-listing acts).

Intertextuality, however, is hylomorphic, too. Compositions have their structure and formal principles of combination; and particular components come together under certain conditions to produce unified entities. Complexities built into plastic and plastic-texts are distinct to each structural formation. The constituents and the construction do not resemble each other and need not be coincident entities. Plastic formations can have an amorphousness and long-windedness, but they are never without the "right" properties of interaction and transaction. This is structured plasticity; in fact, plasticity inheres in the structures within which it exists. Michele Paoletti sees the place of the structure as occupied by each *relatum*.[55] Plastic intertextuality has structures with a certain order—in general, the fact S (A,B,C) is not identical with the fact S (B,C,A)—and breakdown of order, which, again, is relational on its own configural merit. Thoughts and concepts and atoms and molecules have their own places in the intertextual program and process. With places come contiguity, the power and innovative potency of the *relata* and the pattern of the *relata* to express a particular set of meanings. This relation is complex in its distance and proximity to the *relata*. Substructures are formed through *internal* relations that contribute to

---

54. Bakhtin, *The Dialogic Imagination*, 276.
55. Paoletti, "Structures as Relations."

the maximal structure. *The Waste Land* is not a poem about, say, sports or culi-
nary experiences. So the maximal structure forms around certain crystalline
coordinates—postwar sterility, existential hollowness, deficit of regeneration,
etiolation of all aspects of living and living with the world—that are supported
through substructures (for instance, understanding of Christian myths, Hindu
and Buddhist philosophy) having their own internal relations of order and place.
These substructures build their own patterns of assemblage to "bond" with the
maximal understanding. Being a plastic molecule cannot be the text's only ontology; the hylomorphic border pressure brings protean molecularity into our
experiences and thinking of the text.

On the optical dimension of plastic, some are perfectly transparent and
some are translucent and opaque. This variation contributes to the material-
aesthetic where the passage and refraction of light become a critical case in
point. The variation in transparency depends on the microstructure; the "scattering of light occurs from multiphase structures, the phases differing in refractive index and being of a size comparable with the wavelength of light.
Scattering of light and a consequent deterioration in transparency can also result from surface irregularities, again on the scale of the wavelength of light,
which can be introduced during processing."[56] The text has its own levels of
transparency, and reading is refractory depending on the microstructure of the
text and the surface scattering (every polymer-unit, whether "Death by Water"
or "What the Thunder Said," demonstrates its own levels of transparency and
refractive abilities). This is another form of intertextual space where a certain
clutch of properties come together to make a fresh sense of plastic experience
within an individual polymer-text. Some texts are inert, some are resistant to
easy solubility that appropriates the texts from a variety of perspectives and
positions, while some are chemically (re)active with a relatively open structure
that can exist only through inter-domaining. I call this the "chemo-poetics" of
literary understanding. Plastic-aging is like text-aging where the former undergoes reorganization of structure, a lowering of the thresholds of fluidity and
permeability, and material restructuring. It is also dependent on environmental factors such as temperature, oxygen, ozone, moisture, and light, as well as
some biological factors. This aging is ceaseless and not an immediate deci-
mation. These are factors in the change-limits—a kind of transcendence of

---

56. Birley, Heath, and Scott, *Plastics Materials*, 5.

plastic's own existence in relation to time, its history of existence, and external stimuli. Similar complexities engendered through "change limits" infect and inform a text. Texts as heterogeneous as Joyce's *Ulysses*, Juan Goytisolo's *Reivindicación del Conde don Julián*, Salman Rushdie's *Midnight's Children*, Umberto Eco's *The Name of a Rose*, Steven Pressfield's *The Legend of Bagger Vance*, and Laura Esquivel's *Malinche* reorganize themselves on points of permeability, solubility, and aging through chemo-poetical relationship with social, popular-cultural, religious, political, economic, and other factors.

The Poetics of Flow

The rheology of the text is the poetics of flow where deformation comes through stress (force) at different points of the polymeric structure.[57] Textual mobility is ensured as different chains of thought flow past each other. Kirk Kantor and Patrick Watts show that in plastics flow can occur in four different ways: pressure flow, drag flow, shear flow, and elongation flow: "the first two types are driving mechanisms for flow to occur. The last two types describe how the fluid deforms during flow. Stress accompanies the driving forces and deformation is quantified by strain."[58] Pressure and drag matter integrally for the flow to "happen." While we read a text or an event, there is always a pressure flow coming out of pressure differences. A major authorial image or conceptual obsession or motivating symbol make for a high-pressure zone, which initiates a flow into the low-pressure zones in the form of other conceptual and epistemological units or linkages. This process can also extend to the tradition-modernity dialectic where the pressure gradient between the local (which can be loosely read as tradition) and the nonlocal (the centrifugal, uncritically acknowledged as the global) can initiate flow of different forms.

To engage with a different instance from a different continent, we have the Jikken Kōbō[59] (Experimental Workshop) in post-1945 Japanese avant-garde art, which demonstrates this pressure flow and deformation in the plastic-processing of a play. Cross-chain experiments in artistic styles, the use of materials, and the combinatory poetics of tradition and modernity mark

---

57. "Rheology is the branch of physics that deals with the deformation and flow of matter, especially the non-Newtonian flow of liquids and the plastic flow of solids" ("Rheology").

58. Cantor and Watts, "Plastics Processing," 195.

59. Tezuka, "Jikken Kōbō," 171.

the plasticity of this fourteen-member group. Jikken Kōbō, with its eclecti-cism and aversion to permanent categorization, introduced the "plastic princi-ple" into its thinking of art and its relation to changing times and spirit. Western modernism and traditional Japanese culture came into a chain-entanglement involving Cubism, Constructivism, Surrealism, and the Bau-hausian pursuit of total art. Its manifesto calls for a rheological deformation that sees "the possible relations between the kinematics and dynamics" of the text.[60] In the background of "the ambition for the new, virtually and exas-peratingly synonymous with things Western" and "the aspiration for the au-thentic, often symbolized by things Japanese or Asian,"[61] the 1955 play *Pierrot Lunaire* (*tsuki ni tsukareta piero*), an avant-gardist product of Takechi Tetsuji and Kōbō members, polymerized traditional arts with experiments in contemporary time. *Pierrot Lunaire* exhibits shear flow while adjacent fluid movements (here, meant as the intertextual units of understanding) develop different velocities of their own. It is a different polymeric experience emerg-ing out of a friction built through such flow patterns. *Pierrot Lunaire* finds its chains aligned in a particular direction and orientation.

Based on Arnold Schoenberg's 1912 score of the same title, *Pierrot Lunaire* was constructed with "differentials" that lowered the viscosity of the play by opening up more intra- and intertextual spaces: the "void space between the molecules" of ideas and conceptual and performative appropriations. Miwako Tezuka points out how, musically, "Takechi bridged Schoenberg's innovation of *Sprachmelodie* (spoken melody, more commonly known today as *Sprech-stimme*) with a technique of *utai* singing used in Noh plays. Although their origins were a world apart, *Sprachmelodie* and *utai* both consider the human voice as an amalgam of musical instrument and speech device."[62] The molec-ularity of *Pierrot Lunaire* was further revealed by Takechi's "ability to find a common ground in a scientific analysis of human psychology and the nonre-alistic approach of the 'dreamlike Noh' (*yūgen nō*). Takechi was not only con-versant in the aesthetic of traditional theater but also competent in applying the theory of psychoanalysis and Surrealism in his discussion about the the-ater of anti-illusionism in which he found the future of the field."[63] Plastic

---

60. Malkin and Vinogradov, *Rheology of Polymers*, 1.
61. Masatoshi and Reiko, "Readings in Japanese Art after 1945," 380.
62. Tezuka, "Experimentation and Tradition," 73.
63. Tezuka, "Experimentation and Tradition," 73.

process involves a negative correlation between temperature and viscosity: a text does not have a steady "temperature" to itself and is always exposed to a variety of readings and provocations, which render a "temperature fluctuation" in its existence. The shear and elongation flow vary in *Pierrot Lunaire*, which produce a considerable impact on its polymericity and viscosity.[64] As Tezuka points out,

> The mezzo-soprano voice of the *shingeki* actress Hamada achieved the effect of *Sprachmelodie*, whose purpose was to move away from Western musical convention by assimilating the temporal irregularity, the intonation patterns, and the spontaneity of human utterance. Schoenberg's experimental use of the human voice, in turn, proved to have an affinity with *utai*, a vocal accompaniment of Noh performance. Keenly observing the affinity, the Kōbō member Yamaguchi Katsuhiro stated that *Sprachmelodie* would come vividly alive, when contrasted with Takechi's *Pierrot Lunaire* in the abstract visual and gestural style of Noh, just as *utai*, by subtly drifting above the mechanical nature of instrumental music, functions to keep the audience aware of tension between the earthly world and the otherworldly realm of the spirits commonly portrayed in Noh plays.[65]

Caught in such pressure-gradient and flow-deformation, *Pierrot Lunaire* keeps sustaining its energy and flow through "molding." The play manifests planes of interface among reactive agents of thoughts and paradigms—the trans-dialecticism of tradition and contemporaneity—that announce "compatibilization." This brings interfacial adhesion and reduction of interfacial tension among competing polymers. Takechi, through his "theory of denial" (*hitei no ronri*), engineers compatibility and, hence, ensures amorphous textualism. Compatibility guarantees that conceptual units from across traditions and cultures maintain their chains and crosses.

---

64. "In simple shear flow, the vast majority of polymer solutions are pseudoplastic in nature, which means that the viscosity is decreased as the shear rate is increased. The viscosity related to this type of flow is shear-thinning viscosity. Generally, dilute polymer solutions with low molecular weights belong to this category. Another type of flow is elongational, or extensional, flow. In this type of flow, the fluid is stretched—for example, as the fluid flows through a series of pore bodies and pore throats in a porous medium. In such flow, the apparent viscosity is increased as the shear rate is increased. The viscosity related to this type of flow is shear-thickening viscosity" (Sheng, *Modern Chemical-Enhanced Oil Recovery*, 213.

65. Tezuka, "Experimentation and Tradition," 80.

Crystalline-Amorphous Textuality, Continued

In contrast to semi-crystalline polymers, which have both a patterned struc-ture at certain locations and unorganized amorphous regions in them, amor-phous polymers have a loose structure and are predominantly unorganized. A polymer-text shares affiliations with semi-crystallinity and amorphousness to build "mad intertextuality" (as Monika Kaup calls it). This provides "an open exchange between the domain of literature and a 'universe' of intersect-ing scientific, cultural, ideological and literary discourses or 'voices,' and, conversely, the potential rhythm, merging and overlay of those heterogeneous voices within what we usually regard as a 'single literary text.'"[66] How does one get out of a text by being in a text? It is a relationship that "coils" up—amorphous and random but not without the branching heads and ends to con-nect, cover, and curve. A text, as I have argued, runs a temperature—a histori-cal, semiotic, cultural, and metaliterary state of activation and emission—within an equilibrium that is stochastic and sensitive to its own formations and reso-nances. Plastic intertextuality makes possible the knowledge of "what hap-pens" and "what happens to our surprise" and "what happens without our knowledge." Plastic challenges itself and its limits through trans-formations. Here, Umberto Eco makes for a relevant reading, for his intertextual program is labyrinthine and semiotic, questioning the plasticity of interpretation. If get-ting into a text is failing to close the doors on it because compartments start opening and corridors start lighting up, plastic, correspondingly, does not stay confined to simple covalency and inter-chain modes as *incipits* (beginnings) but unmakes itself into an amorphousness with multiple structural bonds and in-tricate overlaps.

The amorphous plastic has its own labyrinths. And plastic intertextuality tries to figure itself out through the three kinds of labyrinths that Umberto Eco mentions:

One is the Greek type, that of Theseus. This labyrinth does not allow any-one to lose his way: you enter it and arrive at the center, and then from the center you make your way to the exit. That is why there is the minotaur at the center; otherwise there would be no point, you would just be out for a

---

66. See Kaup, *Mad Intertextuality*, 12–13.

harmless stroll. The terror comes in because you do not know where you will come out and what the Minotaur will do.

But if you unravel the classical labyrinth, you will find a thread in your hands, Ariadne's thread. The classical labyrinth is its own Ariadne's thread.

Then there is the mannerist labyrinth. If you unravel it, you find in your hands a kind of tree, a root-like structure with many dead ends. There is only one exit, but you can get it wrong. You need an Ariadne's thread to keep from getting lost. This labyrinth is the model of trial-and-error process.

Finally, there is the network, the structure that Deleuze and Guattari call a rhizome. The rhizome is set up so that each path connects to every other one. It has no center, no periphery, and no exit, because it is potentially infinite. Conjectural space is shaped like a rhizome. The labyrinth of my library is a manneristic labyrinth, but the world in which Guglielmo realizes he is living is already structured like a rhizome: that is, it is structurable but never definitely structured.[67]

Among the three labyrinths—the classical, the manneristic, and the network—Eco affiliates with the network, connecting it with the Deleuzian rhizome.[68] This is where the "random coil" polymer model corresponds with the Deleuzian fold, the rhizome—projecting a sort of plastic nomad. The random-coil model "predicts that each long-chain molecule will assume the form of a loose ball, with a diameter that increases as the square root of the number of segments in the molecule. The otherwise vacant space within each coil is taken up by neighboring molecules, and various degrees of entanglement and intertwining are expected not only among segments of different molecules but also among segments of a given molecule. In rather loose but graphic terms, the random-coil structure can be described as similar to an agitated, greatly tangled mass of spaghetti."[69] Within amorphous polymeric material, *sans* external force, each long-chain molecule coils into the shape of a loose ball intertwined

---

67. Rosso and Springer, "A Correspondence with Umberto Eco," 7.

68. "The project of an encyclopedia competence is governed by an underlying metaphysics or by a metaphor (or an allegory): the idea of labyrinth. The utopia of a Porphyrian tree represented the most influential attempt to reduce the labyrinth to a bidimensional tree. But the tree again generated the labyrinth" (Eco, *Semiotics and Philosophy of Language*, 80). It is also important to look at the complete quotation from Borges's "A Note on (toward) Bernard Shaw": "Literature is not exhaustible for the sufficient and simple reason that no single book is. A book is more than a verbal structure or series of verbal structures; it is a dialogue it establishes with its reader and the intonation it imposes upon his voice and the changing and durable images it leaves in his memory. A book is not an isolated being: it is a relationship, an axis of innumerable relationships" (in *Labyrinths*, 213–214).

69. Uhlmann and Kolbeck, "The Microstructure of Polymeric Materials," 96.

with segments of neighboring molecules. The aesthetic potential of the random-coil model connects with Eco's labyrinthine mode. The material-aesthetic here speaks of an asymmetricity—the interpretive capital derived out of amorphous and semi-crystalline polymerization—that lubricates the "turns" and "bends" that it can provide. Plastic synthesis produces its own order; it generates plasticity as asymmetrical chemo-manifestations.

The random-coil, forked paths and the rhizome make possible a different kind of critical consciousness where thinking is "coiling" up with others: a semiosis that builds its own truths and reconstructs truth as well. The plastic literary establishes the event of "open work": it is where appreciation is just not rational but is intuitive too. The event produces the new "chaosmos"[70] as rationality and appropriation constellate to evoke a separate vein of meaning-generation. Such polymerization of meaning speaks of an incompletion of network and growth, coming close to what Eco calls the "aesthetics of chaosmos." Rocco Capozzi notes that the "intratextual and intertextual (hyper)links help us to move from Eco to Aristotle, to Dante, to Kant, to Borges, to Foucault, to Deleuze, etc. etc., and possibly back to Eco, forming an encyclopedic loop of links in a rhizome that the reader can expand according to his own competence. Of course, just as in hyperlinks, one can begin with Eco's work and close with Eco in the same session; or, one can look up other authors (in the/his library) and decide to temporarily set aside Eco's text."[71] Eco's plastic novels are *opera aperta*: protagonists such as Adso, Belbo, William, Casuabon, and Roberto operate and role-play, encouraging frames within frames, folds within folds, texts within texts. They are semi-crystalline and amorphous—for instance, the most plastic of all, *The Name of the Rose*, random-coils "William of Occam, Roger Bacon, Alessandro Manzoni, Jorge L. Borges, Conan Doyle, Mikhail Bakhtin, Charles S. Peirce, Yuri Lotman, Roland Barthes, Maria Corti, Eco's own theoretical and journalistic writings, and so on."[72] Eco invests in bonding, repetition, combinatorics, and branching out more than in originality; the plastic principle manifests in "semiotico ludens" achieved through "inferential walks."[73]

---

70. Joyce, *Finnegan's Wake*, 118.
71. Capozzi, "Intertextuality and the Proliferation of Signs/Knowledge in Eco's Super-Fictions," 466.
72. Capozzi, "Palimpsests and Laughter," 413.
73. Capozzi, "Palimpsests and Laughter," 414.

Plastic intertextuality has alternatives to project and sustain: the "orderly cosmos" and the "fluid and watery chaos."[74] Textual "blanks" demand greater interventionist energies and bonds of interactive chaining. The varieties of material existence and possibilities of manifestation through the impact of external factors help plastic-texts to throw up openness and, hence, radicality—*opera in movimento* ("work in movement" or "in progress"). Plastic intertextuality can be dialectical, interactive, cross-bonding, often unfinished, sometimes conclusive and unavoidable and compulsive. If Eco sees order in Joyce, there is reason to see order in plastic synthesization (symmetry and asymmetry hiding within and promoting each other), too. Eco notes: "I do not believe, however, that the excessive structurality contradicts the open structure of the work. For example, I have always and clearly considered *Ulysses* to be a book in which the model of the open text plays a major role: but *Ulysses* is extremely structured, it has an iron scaffolding."[75] I would argue that this is plastic structure (not iron scaffolding), open and rigid at the same time; it remains plastic and yet transcends its structure through plasticity.

Plasticizer Textuality

Plastic texts exhibit flexibility in response to the temperature of thinking and theory, to the hardness and flex crack resistance (it is a resistance against repetitive strain) and coefficients of friction across multiple thought-genres and conceptual traditions. They exhibit "tear strength" (also called "tear resistance," which is a force against crack propagation); it is here that we have the blessings of plasticizers. Twentieth-century critical consciousness is a sink and space of plasticizers influencing the interpretive life of texts and contributing to its expansion, swell, and resistance. Under plasticization, all major texts that have caught our imagination and built a permanent shelter in the

---

74. Eco, *The Aesthetics of Chaosmos*, 36.
75. Rosso and Springer, "A Correspondence with Umberto Eco," 5. Jessica Pujol Duran observes that "Joyce takes a step further and includes several registers that do not respect the understanding of the whole work as a closed system in the Dantean or medieval sense, but instead include the immense variety of voices and situations with which Bloom interacts through the day. This illustrates the chaotic sense that Eco emphasizes in *Ulysses*. The intertextualities and stylistic varieties that refer to disparate texts and traditions pile up an indefinite sum of parts that may not create a harmonious cosmos or totality" ("Umberto Eco's New Paradigm and Experimentalism in the 1960s," 57).

critical imaginary of the twentieth century are restive and kinetic in their polymer-plasticizer and plasticizer-plasticizer interactions. There are interactions that are transient and ever changing and associations that form, disappear, and then reform. Some texts are visco-elastic—they bend and recover, are unhinged and reshaped—and some undergo longer and relentless deformations. Almost all major texts across world literature through the century are plasticizer-textual in nature, with tensile, compressive, and shear forces to go with them. Plastic texts bear out tensile (the force that enables elasticity) compliance to a tradition—the time-spirit—but are not without tensile stress that sees them expanding into other aesthetic-textual growth with different curves and tangents, elongation and variation.

Pigmentation adds variety and distinctness to plastic; inorganic fillers of silica are added to produce a material change that can make it less expensive, glass fibers are blended to strengthen its formation, and thermal and UV stabilizers are incorporated to make it more heat and light resistant. Additives modify behavior; they increase rigidity and dispersion of particles, and they often, reduce stiffness and bring the composite effect of plastic to a more varied effect. Plasticization that "in general, refers to a change in the thermal and mechanical properties of a given polymer" can be achieved "by mixing a given polymer with a plasticizer or with another polymer (external plasticization)" and "by chemically modifying the original polymer with a comonomer, which reduces crystallinity and increases chain flexibility (internal plasticization). To be effective, the plasticizer must be mixed and fully incorporated into the polymer matrix."[76] The "unit operations" involved are melting, mixing, and homogenization—the plastic shaping is plastic compounding and componentialization. This "additive-performative" makes writing and thinking continuously involved in a series of innovations and movements. On this axis of the material-aesthetic, then, how interesting it is to call Eliot's *The Waste Land* a PVC?

Polyvinylchloride, otherwise known as PVC or vinyl, is one of the most analyzed, tested, and versatile thermoplastics. In fact, "PVC can be rigid or flexible, clear or opaque. It can be processed as a solution, paste, or powder; extruded; injection-, dip-, or roto-molded; knife- or reverse-roll coated. It is easily printed and bonded by heat, radio frequency welding, or solvent."[77]

---

76. Langer et al., *Plasticizers Derived from Post-Consumer PET*, 4.
77. Carroll et al., "Poly(Vinyl Chloride)," 61.

Its variability and elasticity of expression owe to the outstanding correspondence and assimilability of vinyl and plasticizer. In the interactions between PVC and a plasticizer we find a kineticism within the polymeric structure that reduces the glass transition temperature and influences the structure through the length and "branchiness" of side chains, viscosity, and polarity. *The Waste Land* would have been rigid and strict in its meaning-making abilities without the plasticizers. Although there are "impact modifiers" that control the expressive content of the poem, forcing it to stay close to the institutional and the reified, the poem has its own series of lubricants and plasticizers that keep influencing the "fusion properties" of the text, thereby extending, in the process, the threshold meaning as well.

The Waste Land–PVC has its own polymeric status with a particular temperature of concepts and viscosity of discourses to regulate its structure; it has its own polarity and modulus (in the sense of measure of stiffness or elastic stiffness). But a structure comes under revision and reconstruction when plasticizers—both internal and external—are introduced into its state and status, forging change of temperature (lowering the gate-temperature of an idea or concept and hence, allowing allusive ingress), which in turn results in structural mobility and flow. Interactions bring change and with it the equations of compatibility and conformity change as well. *The Waste Land* does not reveal any apparent complication but the complexity stays embedded within the polymeric matrix of the poem. This helps the poem to dechain and rechain itself, producing "free volume" and fusion. Free volume is "a measure of the internal space available within a polymer. As free volume is increased, more space or free volume is provided for molecule or polymer chain movement, making the polymer system more flexible. In the unplasticized polymer, free volume arises from movement of polymer end groups, polymer side chains, and internal polymer motions. Free volume can be increased through modifying the polymer backbone by adding more side chains or end groups or by incorporating smaller molecules with flexible end groups, which can move and rotate." The plasticization contributes to the free volume, and plasticizer molecules are "free to self-associate and to associate with the polymer molecule at certain sites and then disassociate. As these interactions are weak, there is a dynamic exchange process whereby as one plasticizer molecule becomes attached at a site or center it is rapidly dislodged and replaced by another. Different plasticizers will yield different plasticization effects because of the differences in the strengths of the plasticizer-polymer and plasticizer-plasticizer interactions."[78] *The Waste Land* is exposed and invested in different plasticizers, and varying plasticization effects cause the dislodging of meaning-sites through more side chains and end groups of trans-epistemic discourses and references; the poem starts to free up its "meaning" volume. The molecularity changes, and so do the expressive content and potencies. The poem, as it exists, hides the plasticizers, but a close reading reveals the structural complexity and the patterns of co-bonding and co-polymerization.

It is within such a framework that the plastico-poetics—the lubricity theory in particular, as shown in figure 5—of the *Waste Land* uses, for instance, the fable of the thunder from the *Brihadaranyaka Upanishad* as the external plasticizer:

> Then spoke the thunder
> DA
> *Datta:* what have we given?
> My friend, blood shaking my heart
> The awful daring of a moment's surrender
> Which an age of prudence can never retract
> By this, and this only, we have existed
> Which is not to be found in our obituaries
> Or in memories draped by the beneficent spider
> Or under seals broken by the lean solicitor
> In our empty rooms
> DA
> *Dayadhvam:* I have heard the key
> Turn in the door once and turn once only
> We think of the key, each in his prison
> Thinking of the key, each confirms a prison
> Only at nightfall, aethereal rumours
> Revive for a moment a broken Coriolanus
> DA
> *Damyata:* The boat responded
> Gaily, to the hand expert with sail and oar

---

78. Godwin, "Plasticizers," 488.

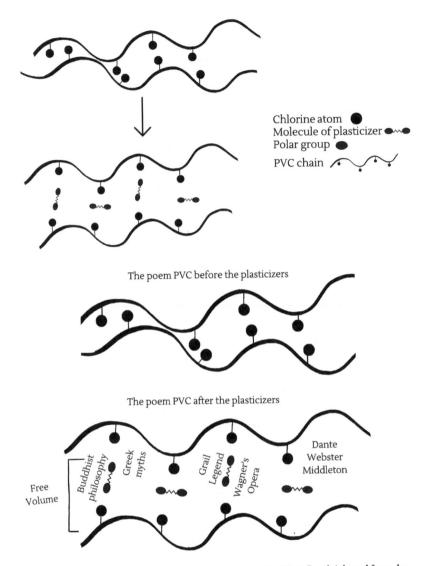

Figure 5. Illustration of lubricity theory in relation to *The Waste Land*. Adapted from the author's drawing by Shreya Pothula.

The sea was calm, your heart would have responded
Gaily, when invited, beating obedient
To controlling hands[79]

Addressing a decaying situation—the decrepitude and dismay—the waste land awaits the arrival of a heroic figure, the ushering in of rain and rejuvenation. Eliot plasticizes the chain reading with the thunder fable from the *Brihadaranyaka Upanishad* in the context of a scenario of the Ganges awaiting rain as thunder rumbles across its face. This external plasticizer has higher polarity than the poem-polymer—as is the case with the working of all plasticizers in general—enabling polymeric loosening of meaning effect. The macro-polymeric chains of the poem describe an apocalyptic situation where Christ is not resurrected and things fall apart, politically and culturally, in the wake of the First World War: cities expire over the dessiccation of culture; they look "unreal." Redemption appears to be out of sight as the "ruined chapel" signals devitalization to the core of our existential being.

It is here that the introduction of the external plasticizer through the *Upanishad* separates the polymer chains and makes the molecules of thought move freely, expanding the "free volume." The thunder fable increases the modulus of elasticity of the polymer-poem. Part V, chapter 2 of *Brihadaranyaka Upanishad* reads as follows:

Prajapati had three kinds of offspring: gods, men and demons (*asuras*). They lived with Prajapati, practicing the vows of brahmacharins. After finishing their term, the gods said to him: "Please instruct us, Sir." To them he uttered the syllable *da* and asked: "Have you understood?" They replied: "We have. You said to us, 'Control yourselves (*damyata*).'" He said: "Yes, you have understood."

Then the men said to him: "Please instruct us, Sir." To them he uttered the same syllable *da* and asked: "Have you understood?" They replied: "We have. You said to us, 'Give (*datta*).'" He said: Yes, you have understood.

Then the demons said to him: "Please instruct us, Sir." To them he uttered the same syllable *da* and asked: "Have you understood?" They replied: "We have. You said to us: 'Be compassionate (*dayadhvam*).'" He said: "Yes, you have understood." That very thing is repeated even today by the heavenly voice, in the form of thunder, as "Da," "Da," "Da," which means: "Control

---

79. Eliot, *The Waste Land*, 18–19.

yourselves," "Give," and "Have compassion." Therefore one should learn these three: self-control, giving, and mercy.[80]

The original plasticizer-text interprets the *da* differently by three orders of existence: gods, men, and demons. Manju Jain observes that this "sequence provides an orderly descent through the scale of existence and gives an indication of the shortcomings of each of the three orders. The fable concludes, however, by exhorting men to practice all the three injunctions for it is suggested that there are no gods or demons other than men. Eliot too sees them as pertaining to the human condition. He altered the original sequence possibly because he believed that the two imperatives of 'give' and 'sympathize' are necessary prerequisites for the attainment of the self-control."[81] The plasticizing of the thunder fable into the texture of the poem achieves a distinctive feat: the mere poetic simplicity of stating sympathy, "giving" and, hence, self-control as *de rigueur* to the emotional and existential desiccation of human situation would have left the poem rigid and strongly bound within the intrinsic forces of a modernist poem. But by importing these concepts through the *Upanishad* with a re-reading and re-living of the fable of the thunder, the introduction of Prajapati as against Dante's *Inferno*, and critical engagements with F. H. Bradley's *Appearance and Reality*, the lubrication within the polymer-text is set in operation. Like PVC, *The Waste Land* has its amorphous and crystalline regions that respond to plasticizers. Portions of the PVC polymer begin to solvate as the interactions growing out of the polymer-plasticizer mix make the product moldable. Lubricity theory of plasticization explains that "plasticizers act as lubricants to facilitate polymer chain movement when a force is applied to the plastic. It starts with the assumption that the unplasticized polymer chains do not move freely because of surface irregularities and van der Waals attractive forces. As the system is heated and mixed, the plasticizer molecules diffuse into the polymer and weaken the polymer-polymer interactions." The lubrication is initiated as the intermolecular forces are reduced resulting in the enhancement of the "flexibility, softness, and elongation of the polymer."[82] The text corresponds closely with lubricity theory, which states that the plasticizer acts as a lubri-

---

80. *Holy Upanishads.*

81. Jain, *A Critical Reading of the Selected Poems of T. S. Eliot,* 189–190.

82. Godwin, "Plasticizers," 488.

cant between polymer molecules. *The Waste Land* as a polymer text is flexed as molecules of thought and ideas graft and work into each other. The *Brihadaranyaka Upanishad* acts as a molecular lubricant allowing the polymer-poem greater mobility of movement and flow; it provides the "gliding planes" and enables the process of "freer movement." It is the plasticizer efficiency that changes the receptive and reactive merit of the text.

Plasticizers "contain polar and nonpolar groups, and their ratio determines the miscibility and compatibility of a plasticizer with the original polymer." The "polar groups present in plasticizer molecules, which usually consist of ester groups, interact with the polar sites of the polymer molecules, namely chlorine atoms, causing PVC chains to spread."[83] The plasticizer is expected to create a stable mixture with the original PVC, which means plasticizers must find stability and "form-correspondence" with the original polymer; like the polarity of the Fisher King, nursery rhymes, Kyd's *The Spanish Tragedy*, the *Upanishad* exceeds the polymer polarity of the poem proper to create the flow and consequent compatibility of the polymer polarity.

It should be mentioned that for "a plasticizer to be effective and useful in PVC, it must contain two types of structural components, polar and apolar. . . . The balance between the polar and non-polar portions of the molecule is essential. If a plasticizer is too polar it will tend to act more as a solvent at room temperature and yield a product with overall poor performance. If it is too nonpolar, compatibility problems can arise with high levels of plasticizer exudation."[84] *The Waste Land* maintains a polar-nonpolar balance through its own plasticizers ranging across critical philosophy, thing-theory, religious studies, eco-critical perspectives, psychoanalysis, Zen philosophy, and others. For instance, one may look into this balance between *The Waste Land* and Henry James's "The Beast in the Jungle." Katherine Spurlock points out that, like *The Waste Land*,

> "The Beast in the Jungle" is a story of epistemology as it focuses on the "horror of waking" to "consciousness." . . . *The Waste Land*, which also is about missed opportunities, buried desires, and sprouting reminiscences, ends similarly by mourning a missed "moment's surrender"—put forward as the only possible redemptive act in "What the Thunder Said." For our purposes, the

83. Langer et al., *Plasticizers Derived from Post-Consumer PET*, 1.
84. Godwin, "Plasticizers," 488.

crucial similarity between the two works is that both invite interpretation as products of their author's psychosexual distress (by-products of James's bachelorhood and Eliot's failing marriage), while at the same time aggressively presenting themselves as self-consciously constructed interpretive puzzles.[85]

Polymer polarity demonstrates how the "compatibility problem" can be overcome. It is the flexibility quotient of the poem that influences the energy within its system impacting on the viscosity and molecular orientation of its expression. The poem-PVC alters its crystallinity with changes in conceptual and contextual temperatures. This alteration affects the transition of every crystalline formation; it also affects the increase and decrease in slope and curve of reception and understanding as we find a variety of plasticizers in the form of Shakespeare's influence on Eliot, Nietzsche and history, the correspondence with Henry James's "The Beast in the Jungle" and many others aggravating the modulus of the poem's textual formations. The efficiency of the plasticizers shows on the meaning-effect of the poem-polymer as the poem demonstrates the efficiency and effect-value of those plasticizers that achieve the desire-effect without intensifying their presence in the changing structure. This efficiency, however, does not reduce the importance of "plasticization threshold," which is a limit that needs to be overcome before the effects of the plasticizer start to manifest on physical properties. Plasticization cannot be without certain points of limit; thresholds matter in all plastic interpretability. *The Waste Land* is no exception.

Within the framework of gel theory (see figure 6), we encounter the stress of internal plasticizers where the polymeric chain dilutes to produce a lot of mobile space, in this case in the last ten lines of the poem. The burst of plasticizers in the end—the figure of the Fisher King (from Jessie Weston's *From Ritual to Romance*); an echo of the words of the prophet Isaiah to King Hezekiah; the prayer of the Italian poet Jacopone da Todi; a refrain from a nursery rhyme ("London bridge is falling down"); Dante's *Purgatorio*; an anonymous Latin poem, "Pervigilium Veneris"; Tennyson's *The Princess*; a line from the sonnet "El Desdichado" by Gerard de Nerval; Thomas Kyd's *The Spanish Tragedy*, and the *Upanishad*, with the Sanskrit word *shantih*—all these "gel" the main polymeric chains that talk about a world in fragments and

---

85. Spurlock, *Rival Authorities*, 221.

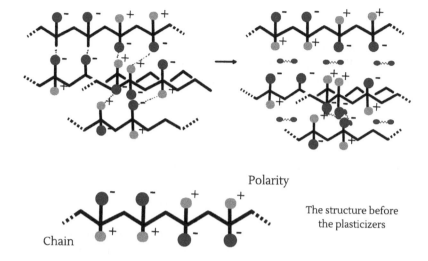

Polarity

The structure before
the plasticizers

Chain

Plasticizers : 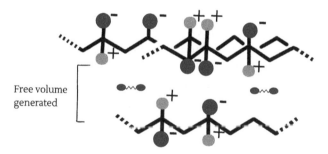 Fisher King, refrain of nursery rhyme, Dante's purgatory, anonymous Latin poem, Tennyson's "The Princess", etc.

Free volume
generated

Figure 6. Illustration of gel theory in relation to *The Waste Land.*
Adapted from the author's drawing by Shreya Pothula.

Eliot in physiological disarray, direly in need of psycho-spiritual *shantih*, consolation, and control:

> I sat upon the shore
> Fishing, with the arid plain behind me
> Shall I at least set my lands in order?
> London Bridge is falling down falling down falling down

Poi s'ascose nel foco che gli affina
Quando fiam uti chelidon—O swallow
Le Prince d'Aquitaine à la tour abolie
These fragments I have shored against my ruins
Why then Ile fit you. Hieronymo's mad againe.
Datta. Dayadhvam. Damyata.
　　Shantih　　shantih　　shantih[86]

Working within gel theory, we see that the plasticized polymer is neither solid nor liquid but an intermediate state, which is

> loosely held together by a three-dimensional network of weak secondary bonding forces. This network or gel could be formed by permanent intermolecular ties or by ties which form in a dynamic state, as plasticizers associate and disassociate with the polymer. In effect the plasticizer provides the role of a bridge between two polymer chains. The bonding forces acting between plasticizer and polymer are easily overcome by applied external stresses allowing the plasticized polymer to flex, elongate, or compress.[87]

The clutch of plasticizers in the end flex and dilate the poem-polymeric chains; the plasticizers referring to Sanskrit, Italian, Latin, and the Spanish literary contexts open up a free volume in border-bashed planetarity, in a kind of cosmopolitan co-habitation. The plasticizers, in fact, bring down the internal resistance and prevent the "reformation of the rigid matrix."[88] The macromolecular thought processes slip over each other to generate the internal space in the poem. The free volume is the product of the motion of the chain ends, the side chains, and the main chain. The poem, through its plasticizers (both external and internal), demonstrates these three kinds of motion. The poem survives on "alternatives" and, in its malleability, invites a habitation in cultural and valuational polysystems. This plasticizer-mesh and their "loosening impact" expose the "productive weakness" of the polymeric structuration of *The Waste Land–PVC*. As part of his poetic art, Eliot deliberately, but also often unintendedly, leaves the attachments of plasticizers to the polymeric chains weak. This results in the chains being separated from

---

86. Eliot, *The Waste Land*, 19–20.
87. Godwin, "Plasticizers," 488.
88. Bocqué et al., "Petro-Based and Bio-Based Plasticizers," 13.

each other, enabling the movement and flow of polymeric [thought] molecules. It generates plastic thinking.

Conceptual plasticizers take control of the polymeric chain of thought (sometimes, although not necessarily, increasing the free volume) and initiate compound processing to effectuate flexibility of performance. In fact,

> plasticizers add flexibility to rigid polymers, rendering them less hard and more resistant to impact (i.e., less brittle). By becoming dissolved and intimately mixed with the long chain molecules of the polymer, plasticizers disrupt the secondary bonds that hold the polymer chains to one another and create more room for chain motion. Successful plasticizers are typically organic compounds that have a lower molecular weight than the host polymer to aid dissolution, yet exhibit low enough volatility to prevent rapid evaporation and loss of the desired effects.[89]

With plasticizers in different forms and variety in the contexts of *The Waste Land–PVC*, derived from across traditions and cultures of thought and use, and from a varied material availability, we encounter softening (accommodative of more spaces for creative appropriation), extenders (supplementation), and lubrication (cross-border epistemic cannibalism). *The Waste Land* has its own "distensibility" through several "ester plasticizers" that can make it reduce its viscosity of reception and expression and lower the elastic modulus of a strict understanding within a modernist aesthetics. The poem's polymeric structure is affected through such additives or dispersants, resulting in an elastomer-text.[90] The weight (conceptual-ideational) of the poem has to keep the poem's movement going and the molecular weight cannot be very high either (poems with very high "molecular weight" refuse to flex and shear). There is a subtlety of chemo-poetics where combinations are nuanced through the right density, temperature fluctuation, and reactive abilities; this process contributes to the proper molecularity of expression. It is what I call

---

89. Csernica and Brown, "Effect of Plasticizers on the Properties of Polystyrene Films," 1526.

90. "Elastomers (rubbers) are special polymers that are very elastic. They are lightly cross-linked and amorphous with a glass transition temperature well below room temperature. They can be envisaged as one very large molecule of macroscopic size. The intermolecular forces between the polymer chains are rather weak. The crosslinks completely suppress irreversible flow but the chains are very flexible at temperatures above the glass transition, and a small force leads to a large deformation" ("Elastomers"). See also McKeen, *The Effect of Sterilization Methods on Plastics and Elastomers*, 305–351.

"esterification" in literary formation that keeps rigorous, yet delicate, reactive groups alive and functional. *The Waste Land* typifies that. Texts, both crystalline and semi-crystalline and amorphous, have their own habits and modes of esterification.

The plastic turn, hence, commits to the "poetics of additives," as we see Oiticica, Schwitters, Eliot, Joyce, Gertrude Stein, Umberto Eco, and others perfecting this art of esterification that keeps changing the properties of their texts and the intrinsic stress and strength of polymeric chains of thinking and expression. In the elastic modulus of these texts, the concentration of conceptual plasticizers brings a distinctive difference. Whether in *Merz* or *The Waste Land*, the "crossover concentration" determines the texture and modulus of the text and lowers the "glass transition temperature" (here, conceptallusion grid transition) to let a new transformation emerge. The product becomes plastic through transition from one state to another. The texts, as elastomers or plastimers, transit from one state to another possible state, declaring their plastic-properties. The connection between concentration and temperature in the material compares with the appropriation and constellation that work in the text aesthetically. In fact, ester plasticizers (say, textual esterification) cannot be applied randomly, for every plasticizer bears out its selection or election on the modes of compatibility with the host material/text. It has to fit into the chain of demand and understanding, marking out its specifications with the dominant properties of the material/text it is being added to. The order in the apparent randomness in *Merz* or Eliot's studious sense of order in intercultural bricoleuring speaks of a concrete technology of writing wherein plasticizers and superplasticizers make a decisive alteration in the workability and performativity of the final mix(text)ture. The efficacious and fecund outgrowth of acts owe to a method in the poetics of additives. Texts are ontologically plastic; plasticizers through order, method, experimentation, and transformability make and keep them performatively plastic.

Coda

How aesthetic, then, are plastic molecules? How *articulate* are polymers? Reflecting on the molecule, Roald Hoffmann observes:

It seems there's nothing beautiful in its involuted curves, no apparent order in its tight complexity. It looks like a clump of pasta congealed from primordial soup or a tapeworm quadrille. The molecule's shape and function are enigmatic (until we know what it is!). It is not beautifully simple. Complexity poses problems in any aesthetic, that of the visual arts and music as well as chemistry. There are times when the Zeitgeist seems to crave fussy detail-Victorian times, the rococo. Such periods alternate with ones in which the plain is valued. Deep down, the complex and the simple coexist and reinforce each other.... The world of real phenomena is intricate, the underlying principles simpler, if not as simple as our naive minds imagine them to be. But perhaps chemistry, the central science, is different, for in it complexity is central. I call it simply richness, the realm of the possible.[91]

Intricacy, the limits of the possible, and richness potentialize a plastic molecule with *techne* and *aisthe* in the sense of art and perception (the agency of the perceiver). The molecule is technical-material; it is aesthetic-ethical also. If modern chemistry "is exactly the art (*techne*) that provides creative access to what Plato considered the realm of the most beautiful bodies," making up "the four antique elements" like "fire/tetrahedra, air/octahedra, water/icosahedra, and earth/cubes,"[92] the aesthetic translates to order, plan, and performance in the realm of thinking and action. The plastic-molecule is inherently beautiful, complex, arty, performing, enigmatic, and always in motion and vibration; plastic texts are caught in similar congeries of performatives.

The molecule models (for instance, in spherulite, random-coil, or semicrystalline structures) and builds an interesting connection between its structure and epistemic-literary-cultural understanding. The plastic literary cannot be without form, and it is through form that it feels and experiences plasticity. Polymeric intertextuality, as all these analyses demonstrate, becomes, in the words of Eco, a "play between the appearance of order and the suspicion that there is no order, or rather . . . that there are many kinds of order, and that all of them must be tried in order to reach some (provisional) solution."[93] Plastic discovers and settles into its plasticity—the plastic-order—through a design, an ability to synthetize and shape at the same time, and

---

91. Hoffmann, "Molecular Beauty," 197.
92. Schummer, "Aesthetics of Chemical Products," 84.
93. Rosso and Springer, "A Correspondence with Umberto Eco," 6.

an articulate power to speak about and beyond its immediate modeling. The plastic literary works through forms that are often resistant and rigid, on occasions immune to corrosion and molding, and at certain times malleable and fluid; it is performative-pedagogic.

Bernadette Vincent points out that "because the mechanical properties of heterogeneous structures depend upon the quality of interface between the fiber and the polymer, it was crucial to develop additive substances favoring chemical bonds between glass and resin. The study of interfaces and surfaces consequently became a prime concern, and gradually reinforced plastics gave way to the general concept of composite material."[94] This concept of the "composite"—the bringing together of different phases of structuration and synthetic processes—underlies plastic intertextuality further. Composite materialization, with its own standardization and substantiation, as distinguished from ordinary consumer plastic, can become "high-tech," artistic, and innovative. The composite introduces its own avant-gardist potencies. Composite plastics thus allow the interaction of four variables—structure, properties, performances, and processes—to initiate new approaches and styles, designs and desires. Both plastic intertextuality and avant-gardism are not without these interacting variables.

The textuality of plastic and the plasticity of text avail themselves to be rethought and re-performed. The plastic molecule, like the workings of the plastic literary, invites experimentation with a variety of designs and models; it promotes norms and principles of combination and coherence to produce fresh realities of an ever-mattering material. Plastic composite yields productively to its plasticity. It knows and discovers the points of resistance, transcendence, and a dialectic of emancipation through self-experimentation. The innate and manifest habits of plastic invite intelligence and innovation and refuse to stay arrested to a specific form of thinking and understanding. Chemo-poetics thus declare the plastic literary as a bundle: the bundle is a structure, "structuring structure," the coming to structure, substantialist ontology, and substantialization. The bundle brings us to think about supramolecular chemistry, which is concerned "with 'soft bonds' that make its entities 'thermodynamically less stable, kinetically more labile, and dynamically more flexible than molecules.'"[95] The plastic literary is always thermodynami-

---

94. Bensaude-Vincent, "Plastics, Materials, and Dreams of Dematerialization," 21.
95. Weininger, "Butlerov's Vision," 155.

cally active; it comes with "self-assembly," "self-organization," and "self-recognition." It makes us realize that critical understanding in general—for instance, of the poverty in South Asia or economic undergrowth or racial escalation or technological globalization or world literature—is the supramolecular bundle but not without its points of structural understanding, points of meaningful agreement, and what we may call an interpretive equilibrium. The bundle as entanglement is dissipative energy and equilibrium in thinking, stability, and vibration. The plastic literary aims for form (the "target molecule") and tries to reach a steady state only as a limiting paradigm before it gets stochastic and restructures itself into another shape and reality.

The plastic turn has produced its own varieties of composite textuality. And the hetero, crystalline-amorphous, and plasticizer textualism clearly show that "the symmetry properties of any object are not mutually independent—they generally come in bundles."[96] Meireles's *Malhas da liberdade* or *The Waste Land* or *Pierrot Lunaire* are bundles that demonstrate aesthetic-chemical symmetry. Joseph Earley argues:

> The geometric structure of each chemical molecular entity is far from a rigid arrangement of atoms in space. Stable chemical species are those that persist in a specific pattern of internal vibratory motion so that average distances and angles are reasonably well defined. . . . The identity of chemical individuals does not depend on single fixed static structure, but rather on the fact that continual return to prior arrangements of parts occurs. . . . Every chemical entity, insofar as it persists in time, exemplifies some group theoretical structure, corresponding to the collection of spatial symmetry properties of the average arrangement of parts that persists through thermal motion. It is in virtue of the closure of relationships that grounds the dynamic stability of such spatial structures that molecules may act as units in interactions with the rest of the world.[97]

Plastic molecularity with its symmetries and asymmetries leads to a bundle that establishes a structure. The constructive in the plastic literary is dynamic equilibrium. Plastic texts, material-aesthetically, project and produce models. Plastic *figures*.

---

96. Earley, "Some Philosophical Implications of Chemical Symmetry," 207.
97. Earley, "Some Philosophical Implications of Chemical Symmetry," 215.

Chapter 3

Plastic Touch

Plastic has provoked and triggered our imagination; it has challenged our imagination as well. What we do with plastic is also what we cannot think of doing with it. Plastic, in its historical and affective ways, builds both transparency and opacity: the opacity comes from the unknown and unrealized potencies of the plastic mattering. The opacity of plastic is our surrender to the limits of representation—a capitulation to the enigma of existence where incommensurability with plastic is about assuring oneself of repeated re-turns and retrenchments in it.

On this note of countertextuality in representation, the material-aesthetic brings a turn to plastic art—art *with* plastic—through a separate vein of aesthetic imagination, creative innovation, and distinct affect-power. Plastic art premises the material-aesthetic through a new dynamic where humanity's relation with plastic goes beyond consumerism, cultural habituality, economic viability, the eco-catastrophic mandate, and spectrality into the (im)pure realm of art and imagination. It becomes a fresh "poetics of connection" that is invested in aesthetic emotion and spatialization. The plasticity of plastic and the plasticity of an artistic mind come into a compelling and convulsive interplay. As Paul Ardenne notes, "Poetic-aesthetic offering is not only visual, a variety of retinal art, nor is it only intellectual, a form of conceptual art; but it is, in addition, and very often, multi-sensorial."[1] Within the laboratory, plastic has been desired into a variety of states; it keeps forming its own desires, too. And, outside the laboratory, it does not cease manifesting in varied shapes and innovative desires. This continues further in plastics' discarded status as weathered, washed up, worn-out versions of themselves, starting to express their desirous ways by evolving and emerging at known and unknown places and igniting plastic imagination through a complicated interface with anthropogenic art-forms and art-meaning. This desire comes from plastics as ready-made, found, waste, and obscure objects. Plastic arts have a desire that is sourced out of an irrationalism and "anti-ocularism"[2]— the surprise is on the artist, and the artist is being spoken back to by the object. A vision is raised in a subject-object negotiation of control and often the lack of it; it also emerges through the states of the seen and unseen and other perspectives that are shaped and constructed through the object. Constructed for human use, plastic-objects come back for another round of use in the hands of the artists: as objects from the laboratory of the earth, with a different valence, serendipity, and use-value. Often, it is the object-desire that shapes and motors the artist-desire and, eventually, the art-desire.

Plastic affect (from the Latin *affectus*, to afflict or to touch) has brought us before a plastic sensorium.[3] The turn to plastic in arts is multisensorial, dialectizing the optic-haptic, the imaginative-analytic, and the aesthetic-didactic. The Romantic gothic in plastic accumulation through uncanny shapes and forms and the discarded plastic through representational art brings the "negative" in heightened intensity, the negative as discharged through anxiety, horror, and the terrible. Freud calls this the "unpleasure." The aesthetic pleasure is weak as plastic arts impact on our psyche with an excitation that makes us see beyond the immediate image. The sublimation achieved therefore holds us in the sway of the negative. Silvan Tomkins sees interest as more powerful than "drives": interest is more motivational and inspirational than drives in the first place; before long drives start to emerge

---

1. Ardenne, "Ecological Art—Origins, Reality, Becoming," 62.
2. See Railton, "That Obscure Object, Desire."
3. Doss, "Affect."

out of interest.[4] Susan Best explains that "in regard to art, Tomkins's theory of affects displaces the Freudian emphasis on the drives as the controlling but silent partners in cultural life and, in the case of art, this allows a more straightforward relationship between positive affect and its object."[5] For Tomkins, this is the interest-excitement. Best notes that "interest is a liking for the challenge of the unfamiliar." It is "about a kind of psychic stretching, perhaps even a restlessness about things as they are, things known."[6] The turn to plastic constructs an interested "communion" with the familiar and, often, strange objects. All the plastic artists discussed in the following pages have a serious corporeal and affective investment to make in collecting, storing, assorting, collaging, collocating, and assembling plastic as every object builds its own desire in an aesthetic reanimation and eco-cultural resurrection. From the ocean to the gallery involves a "travel," a separate sense of multisensorial rhythm and a transfiguration of the debased, deceased, discarded, decrepit, and downgraded line of objects. It builds its own figuration and rationale of taste. Plastic evolves through art with another vein of fascination, as a dazzling product of imagination, but not without its settlement in a moral and psychological torpor. The contrapuntality of the material-aesthetic fuses interest, drive, excitement, and enjoyment.[7]

Gaming with plastic is deeply affective at various stages of the art-growth where both the subject and the object are at play; it is where the artist essentially has no control over objects' supply-order and variety of availability (playful objects) as much as the subject makes the objects *play* to the artist's design. The complexity of identification opens up affective affordances, when, for instance, Finnish artist Tuula Närhinen makes her *Frutti di Mare* creatures refuse a complete assimilation of subject, art, and art objects into an aesthetic whole. Her *Frutti Di Mare* is an installation of thirty-six c-prints on aluminum and thirty-six floating sculptures in plastic cases partly filled

---

4. See the Tomkins, *Affect Imagery Consciousness*, chap. 10, on "Role of Interest." Tomkins notes: "Interest, then, supports the operation and development of a variety of sub-systems—the drives, perceptual, cognitive, and motoric apparatuses, as well as their organization into central assemblies, and governing Images which exploit the human feedback mechanism in the interest of the individual as a whole. We have said that interest is a support of the necessary and the possible. By this we mean that, because interest is linked in different ways to each of the sub-systems, it subserves different functions in the total economy of the human being which range from the necessary to the possible" (189).

5. Best, "Rethinking Visual Pleasure," 505.

6. Best, "Rethinking Visual Pleasure," 510.

7. See Demos, *Exploring Affect*.

with water. These floating sculptures are made of plastic waste and represent "new kinds of marine species from a previously unknown sea called Plastic Ocean."[8] They appear like submerged creatures finally in emergence in our plastic nature when nonplastic creatures have gone extinct or are on their way out. The plastic substitutes have arrived. Närhinen observes, "While working on the Frutti di Mare creatures, I observed how plastic materials behaved in an aquarium. Some of the plastic floated, and while the heavier parts sank down towards the bottom. I started to wonder what would happen if they were released to swim in natural waters?"[9]

It is a "suspension" that informs the artist, art-object, and artwork, affectively and psycho-corporeally. Plastic arts, as much as plastic objects, build a suspension in our aesthetic understanding and judgement. It is this suspension that keeps renewing our interest to turn to plastic, building the libidinal energic cathexis.

"As a visual artist," Närhinen says, "I am possessed by the desire to take the pulse of the seashore. The fascination resides in the process itself, in the oscillation between romanticism and rationalism, or *natura naturans* and *naturata*. Time after time, I return to the seashore and find myself lured by the enchanted voices of the Sirens. I want to dive into the brackish water of inspiration—while trying to stay resilient and keep my feet dry."[10] The nexus between desire and collection informs the Australian artist John Dahlsen's aesthetic program, too. He writes that

Initially my foray into making environmentally based art was entirely accidental, I was collecting driftwood from very remote beaches in East Gippsland in Australia, to make furniture for our new home in Byron Bay and I stumbled across huge amounts of plastic debris that had washed up on the beach and felt compelled to collect it, intending to take it to the local dump for recycling. The more plastic I collected the more intrigued I became by this stuff and shipped sixty to eighty jumbo garbage bags of it back to Byron Bay and left it on the side of the studio while I made the driftwood furniture. Then I poured it all out onto the floor, the colours, the forms amazed me. I was on fire, just with the possibilities of what I had discovered.[11]

---

8. Närhinen, *Frutti di Mare*.
9. Närhinen, "Between the City and the Deep Sea," 158.
10. Närhinen, "Between the City and the Deep Sea," 160.
11. Dahlsen, "Advice from the Ocean."

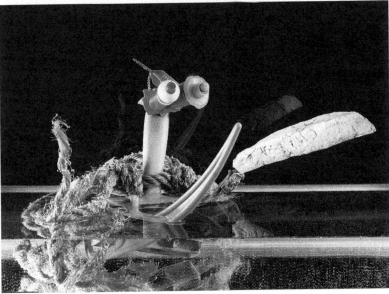

Figure 7. Sculptures from *Frutti di Mare*, an installation by Tuula Närhinen, 2008.
Used by permission of the artist.

The American artist, Tess Felix, uses plastic litter combed from the beach to produce collage portraits of family and friends; she calls these "eco (ocean) heroes." Her art narrates both the "stories" of her subjects and how plastic itself is experienced as a force and an insinuation into our lives. Felix notes, "These ocean plastics convey a long story of their journey starting with extraction of oil from the ground. Plastic is everywhere in our lives, in fact we touch it more than we touch our loved ones—it's just a brute fact that plastic is a 'thing' in the world."[12]

The material-aesthetic comes alive with the plastic artist as a collector. Plastic art corresponds with plastic collector. It is a collection that builds a home for art—a poetics of plastic objects. This reminds us of André Breton's collection of more than fifteen thousand items that included "manmade and natural objects, books, manuscripts and other miscellaneous curiosities and ephemera."[13] It is a deeply complex configuration with the material and the aesthetic. Breton, within the surrealist tradition, and Närhinen, Dahlsen, Felix, Judith Selby, Richard Lang, Evelyn Rydz, and the other plastic artists all have a vigorous investment in "collection." The act of collection—both *naturalia* and *artificilia*—has its own long tradition and history going back to the premodern era. Whether the act of collection can be artistic or degenerative or a product of rationality and methodology are matters for debate. As Christina Helena Rudosky points out, "Balzac represented the collector as a decadent fetishist (*Cousin Pons*), Proust saw the collector as a failed artist (*À la recherche du temps perdu*) and Anatole France deemed the collector a hermetic archivist (*Le crime de Sylvestre Bonnard*)" but "Champfleury interpreted the role of the collector as a radical within his novel, *La mascarade de la vie parisienne* through the character of Topino (a rag-picker and therefore degenerate) who collects posters and interprets them together as one would a collage."[14] Walter Benjamin transforms the notion of collection through his waste-aesthetics, and Breton brings a separate dimension to material-poetics through his atelier collection. Breton chose the bizarre and the marvelous, "valuing the mysterious over the understood and the Baroque over the Enlightened."[15] He brought his own design, intention, and practice into the collection.

---

12. Felix in "Tess Felix: Curious Remnants."
13. Rudosky, "Breton the Collector," iii.
14. Rudosky, "Breton the Collector," 5.
15. Rudosky, "Breton the Collector," 6.

For Närhinen, as for other plastic artists, the act and art of collection are investments in mono-materialism wherein plastic in varied shapes and forms fill up the repository. "Looking at the mess on the seashore," she notes,

> I felt a strong urge to pick up every piece of plastic and sort it out to create some order in the chaos. But my attempt was doomed to fail. The heaps along the shoreline were awash with plastic crumbles: everything from tiny nurdles [tiny pellets of plastic] and unrecognizable broken fragments to intact small items lay thickly packed and intertwined with natural materials in over-whelming layers of debris. The deeper I dug, the more plastic I found. There was no end to my undertaking. The process of gathering was reminiscent of my experiences in picking blueberries or mushrooms. As in the hunt for de-licious yellow chanterelles hiding in the forest, my eye became gradually tuned to even the smallest pieces of colorful plastic.[16]

Breton's choice of material is wide and diverse; but the plastic artists are inveterately plastic-premised, circled within a fixed material choice and del-uged by its abundance as seashores become a collector's cornucopia. Breton brings forth an artistic tradition working within certain aesthetic and philo-sophical principles with separate cultural valence and value to present and promote; these artists commit to an artform that figures the relentless disin-tegrative plastic-effect on Earth and biotic life with a plastic-affect involving emotions ranging from wonder and imagination to dismay and despair. This artform is material-mediality achieved through a form of alienation that in-terrupts our ways of thinking about the world and the varieties of capital that hedge us in. Plastic conjoins through alienation—the practice and po-etics are rendered complicated through attachment, withdrawal, rejection, and, again, a life through the deceased.

The affect of plastic art builds on three coordinates of compelling comple-mentarity: the "out of place" plastic hurtled and washed up from places that can no longer be identified and traced; the place and context it finds within an aesthetic space; and, how it begins to speak in the company of fellow plastics through Bretonian *détournement*. Plastics come together into a fresh context of meaning to construct narratives of varying concerns and desires, intentions and insights. The historical indefiniteness of plastic materials gives way to a

---

16. Närhinen, "True Colours of Twilight," 120–121.

history of the present where the totality of the aesthetic moment declares an autonomy for itself. If Breton, in his surrealist drive, is obsessively devoted to ethnographic objects, Närhinen, as much as Judith Selby Lang and Richard Lang and others, chooses to stay confined to plastic, for the supply is inexorable and inscrutably endless and heterogeneous. Breton's collection is not typed or categorized within a particular range of interest; the supply comes from "experience"—much as it does for the plastic artists—and derives art-materials from miscellaneous encounters. The use-value of objects is never predetermined when one goes about picking and collecting with no clear artistic aim in mind—the material playing up, rather, gaming, the aesthetic, as it were. Plastic nurdles keep heaping up on Närhinen, Selby, and Lang as they struggle to design within a collectorial urgency and dynamicity. Närhinen observes that the "collecting acted like a drug" on her:

> I spent hours scanning the flotsam with keen eyes, my hands eagerly catching every piece of plastic and storing them carefully like berries in a bucket. Oddly enough, digging in the heaps of waste turned out to be addictive to the point that at times I found it hard to stop. Was it an inborn survival instinct or some sort of a primitive drive inherited from the ancient hunter-gatherers searching for edible plants and sweet berries? But this was not about food. I was not even aware of what I was picking up, it was the colors I was after.[17]

Each object comes with an autonomous value of its own as the plastic artists, gradually, through artistic election and collocation, invention and composition, develop a mediatory value among the materials. The artistic and aesthetic construction begin here; it is at this point that the artists start to speak and communicate with their materials. Every single plastic actant gets a voice to speak back to Närhinen, Lang and Selby Lang, Felix, and Dahlsen, and each artist *connects* the actants—in their monadic and conative potencies—within their plastiscapes.

Närhinen's *Baltic Sea Plastique* is distinctly close to my material-aesthetic interests as she provides "analytical drawings" in a series of neatly lined and sketched out plastic entanglements (figure 8). They show an affective and constructive connection between the mind that creates and the mind that

---

17. Närhinen, "True Colours of Twilight," 120.

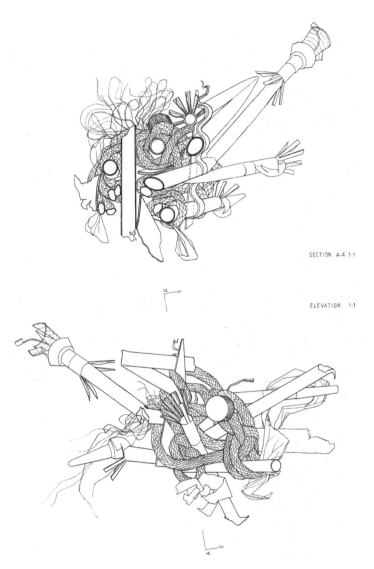

SECTION A-A' 1:1

ELEVATION 1:1

Figure 8. *Baltic Sea Plastique: Analytical Drawings (Coral 1)* by Tuula Närhinen, 2013. Used by permission of the artist.

suffers–a mediation between the design (aesthetic) and the material. She explains that the work presents as

> a kind of closed-loop recycling of the creative design process. New plastic objects start as dreams on a designer's drawing table. Your mobile phone (used for one year), your shampoo bottle (used for one month) or your grocery bag (used for one hour)—are all produced in factories where the raw material, that is the nurdles, gets extruded or molded into design items that maintain their polymeric chains and structural properties for hundreds of years. After being consumed, the products are discarded as trash and—sooner or later—they might end up in the ocean. When the plastic debris floats to shore, it is picked up by an artist who puts together the bits and pieces. Within this work the plasticity of the plastic has run full circle. The analytical drawings return the plastic designs back to the drawing table—the place of origin from where it all started.[18]

Plastic *designs* itself—rather, it comes designed through a poesis of degradation and weathering (aesthetic through the material); artists re-design (the material through an aesthetic). Both connect as different versions of the "analytical." Plastic used and plastic discarded conjoin through manifestations that are creativogenic. Plasticogenically, the abandoned are restored, installed, and re-aestheticized; discard poetics is made to work through experience and experimentality. Distinguished from marine science, this connection with the sea is of a different order, affect, and imaginative energy.

Närhinen characterizes her *Plastic Horizon* as "cultural archaeology" conjugated with "personal obsession." It is a part of her work *Impressions Plastiques*, built through marine plastic waste drifting ashore on Harakka Island in Helsinki (figure 9). Also part of this work are two brilliant tapestries, *Soleil Levant* and *Soleil Couchant*, woven out of discarded plastic bags (figure 10).

Plastic waste is covert and escapist: running away from the producer before long returning in sites and forms that are distinctly unhomely. They accumulate in places and through times that are mostly undetermined, building their own affective potencies. Earth time is threatened to be measured

---

18. Närhinen, "Between the City and the Deep Sea," 159.

Figure 9. *Plastic Horizon*, from *Impressions Plastiques*, an installation by Tuula Närhinen, 2016. Used by permission of the artist.

Figure 10. *Soleil Levant* and *Soleil Couchant*, from *Impressions Plastiques*, an installation by Tuula Närhinen, 2016. Used by permission of the artist.

by plastic shifts—let's say, plastic trash. If plastic is produced to be trashed eventually, this trash, on most occasions, refuses its own obsolescence. William Rathje and Cullen Murphy point out that the "basic methods of garbage disposal are four: dumping it, burning it, turning it into something that can be useful (recycling), and minimizing the volume of material goods—future garbage—that come into existence (this last is known technically in the garbage field as 'source reduction'). Any civilization of any complexity has used all four procedures simultaneously to one degree or another."[19]

The material-aesthetic adds a fifth dimension to the waste-performance: plastic trash into plastic art. In fact, waste is essential to life; screening off the "excess" makes for the cycle of living. Bodily waste and discarded artifacts, notes Joshua Ozias Reno, "share more than symbolic relevance; they actively resemble each other because of the similar interpretive fate they face when separated from the form of life—the living process—that gave rise to them. The transience of decomposing and deteriorating matter can be seen as loss, but also as the perpetuation of life."[20] Interpreted and represented differently, plastic dump renders a fresh order and meaning to waste. The "anxious object" that Harold Rosenberg pointed out persists with a different tenor and affect in plastic-waste poetics working through a complex relationship between the artist, art, and the designing of the project.[21] The plastic bags, toothbrushes, umbrella handles, combs, and other detritus speak of a form of design; it is art *as* design, with a distinct communicative affect and radicality. For Närhinen, the "plastic bag-art" in *Soleil Levant* and *Soleil Couchant* is a dance with the "subject-object duality" (in the words of Timothy Morton). She is a part of the "ecological desire" that does not make nature a simple "mirror of our mind" but admits her to the "unnaturalness both of the object and of the subject," an experience in "radical non-identity."[22]

Within Nicolas Bourriaud's "altermodernism," plastic detritus is the radicant that is defined "through a plant metaphor (ivy, which puts out runners directed by chance that root in haphazard, unpredictable, and often temporary ways)."[23] Närhinen's waste-artification speaks about a centered artist subject and an "unmoored subjectivity." Both the *Soleil Levant* and *Soleil Couchant* demonstrate the radicant in plastic matter, which elicits and solicits spectrality, participation mystique, and a politics beyond our everyday thinking. It is the unnaturalness in the object and the subject where the sun rising at the plastic sea horizon produces a synchronic space that is multiple-voiced in artistic subjectivities and intricate materialization of discarded plastic

19. Rathje and Murphy, *Rubbish*, 33.

20. Reno, "Toward a New Theory of Waste," 9.

21. See Rosenberg, *The Anxious Object*.

22. Morton, *Ecology without Nature*, 85–86.

23. Amy J. Elias, "The Dialogical Avant-Garde," 744. For more on "altermodernism," see Bourriaud, *Altermodern*.

matter. *Soleil Levant* and *Soleil Couchant* are art events in plastic-erotics, in-formed by a transgressive desire, a trans-formative energy, and an aesthetic engagement in wonder and radicality. Plastic waste-art submits to a "refine-ment" that John Scanlan sees as an "effect of differentiation" and a "distilla-tion" that "separates a domain of aesthetic existence from the rational utility that otherwise orders the objects of the world: refinement organizes experi-ence into the graceful and the gauche, the tasteful and the kitsch."[24] The material-aesthetic, through plastic trash, draws a fresh valence from the "sen-sible," separating itself from the rage of the random and the nausea of non-sense: plastic-molding *happens* differently again—differentiation as plasticity.

There is an activism to rise above the "ugly," notes Tess Felix; it is about aesthetic emancipation from the hegemony of the rubbish. This arranging of the ugly "into a beautiful thing" is a demonstration of "the power to over-come the very thing that threatens us."[25] For John Dahlsen, plastic arts re-frame "rubbish ecology"[26] by transforming rubbish into "objects of value." Further, he notes, by "raising questions about the assignation of cultural worth, they also compel the viewer to make links between the cycles of pro-duction and use of everyday functional objects, and those of art."[27] On ma-terializing trash into an aesthetic, Dahlsen observes:

> The plastic garbage was tipped onto the studio floor and it looked to be a po-tential palette. The process of sorting began with the red plastics going in one corner, the blues in the other corner, the pinks somewhere else, the yel-lows in another and then the blacks and so on. Realizing that actually a pal-ette was forming, so many artistic sensibilities started emerging and ideas of working with it, how to actually go about and create something aesthetically beautiful out of it all while working with a challenging medium with its in-herent activist and environmental overtones.[28]

The coming together of plastic material into an act of expression con-structs an aesthetic emotion that is not self-sufficient in itself. In a process

24. Scanlan, *On Garbage*, 89.
25. "Tess Felix: Curious Remnants."
26. Yaeger, "The Death of Nature and the Apotheosis of Trash."
27. Dahlsen, "Artist Statement 2019."
28. Dahlsen, "Painting with an Environmental Palette." See also Dahlsen, "Environmental Art."

where plastic materials, scintillatingly, transform into plastic art-material to "figure" out the plastic art-work, the aesthetic emotion is diffractive. Plastic waste-art thrives in such formabilities, form-possibilities.

Breton's Oceanic objects are subjects that have a narrative to recount; they have a textuality to themselves. All the plastic-artists have an epiphanous and analytic relationship with every nurdle, pellet, and bit of plastic ephemera: a delight to be caught among a jouissance of objects. The plastic turn triggers unremitting affective and interpretive returns, as it spreads out through a riveting and rambunctious network of signifiers and signifieds, putting a compelling plastic semiotics to work. The plastic objects that make their way into artforms have a past-life to themselves—an encrypted story to their present. Ghostly in their objecthood and object-presence, plastics—aestheticized into art—bring a separate worldview as distinguished from their earlier factory-processed and economically determined status. Calling their find titled *Skeleton* "spooky" (see figure 11), Richard Lang and Judith Selby Lang speculate that the object, "may be from a gum-ball machine or a Cracker Jack box, picked out of the rafts of plastic on the beach. As a toy it holds the fascination we all have for death and decay. And it is why we love the Day of the Dead celebrations—to scare ourselves in the reminder of our end. This example is tough to parse with the pelvis somehow distorted but it holds the mnemonic of ribs and spine."[29]

Regarding another piece ironically titled *Love Hand* (figure 12), Lang and Selby Lang write:

> Judith found a doll's hand with the two middle fingers pressed against the palm, index and pinky raised like horns, metal band style. . . . Heavy Metal baby doll? The horns of Satan? The Devil is indeed at work spoiling the world with his crapulous plastic. At home we did our usual web search and found the rock-n-roll sign but the thumb is holding down the two middle fingers. This little pudgy hand has the thumb out hitchhiker style. Oh! Of course it's the ASL, American Sign-Language hand gesture for *I love you*. On the web we found a doll called the "Richard" with that very same hand gesture. Oh, sweet, lucky little love hand pointing the direction on a day in May.[30]

---

29. Lang and Selby Lang, *Skeleton*.
30. Lang and Selby Lang, *Love Hand*.

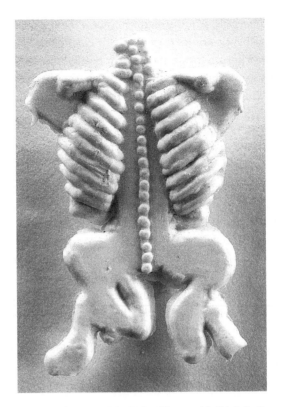

Figure 11. *Skeleton* (2015) by Richard Lang and Judith Selby Lang.
Used by permission of the artists.

Many of Selby and Lang's "arrangements" including *Bottle Nozzles, Fish Handles*, and *Spiff* declare Freudian "latencies"—stories yet untold. Plastic loves to hide. The plastics collected in art installations keep the source-point concealed from the artist as they light up the artist's imaginary with their present and presence haunting a past whose stories we might never know but can possibly speculate and record. For instance, regarding the fascinating *Dinosaur* and *Lamb* (figures 13 and 14), Selby and Lang see the figures as

> two toy animals found in the sand—predator and prey. One is a Tyrannosaur and one a lamb. Ask any 4-year-old what is their favorite animal and most likely it's a dinosaur—safely extinct but holding the fantasy of power. For sweet passivity, the lamb has been in place for millennia. The first toy animals (some with wheels) were often depicting lambs, cows and horses.

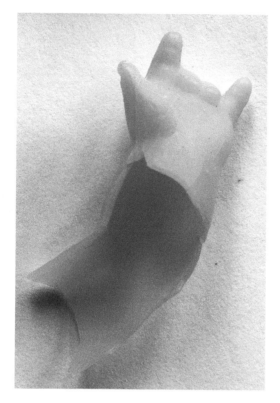

Figure 12. *Love Hand* **(2015) by Richard Lang and Judith Selby Lang.**
Used by permission of the artists.

They date to 4000 BCE but small sculptures of animals go all the way back
to the early stone age, some date to 35,000 years ago. Our minds are ignited
by holding tiny effigies of our animal cousins.[31]

All the plastic-art on display demonstrates a Breton-like attachment to the
objects—a kind of "talking" that they *do* and *perform* by trying to live in a reality that is disturbing. Through *Spooky*, *Love Hand*, = *Dinosaur* and *Lamb* we
find a vibrant *intimacy* and "presencing" with plastic. Lang and Selby Lang
consciously avoid "making" art that resembles a fish or a bird or a portrait or
landscape. In line with the abstract notions of the Bauhaus movement, they

---

31. Lang and Selby Lang, *Dinosaur* and *Lamb*.

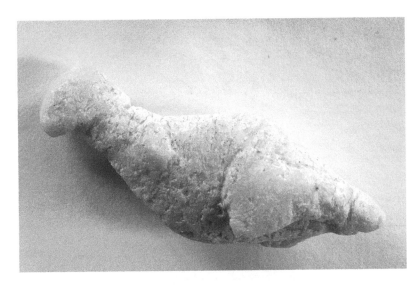

Figure 13. *Dinosaur* (2015) by Richard Lang and Judith Selby Lang.
Used by permission of the artists.

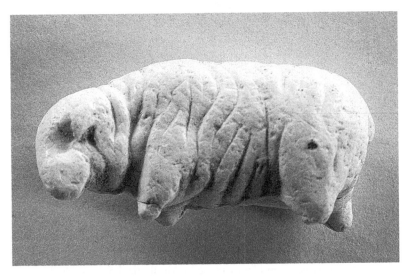

Figure 14. *Lamb* (2015) by Richard Lang and Judith Selby Lang.
Used by permission of the artists.

work through "ideas of rhythm and line, shape and color"—a kind of musical composition. They point out that "the plastic is seen only for color and form, not the things they once were. They become akin to strokes of paint coming off a palette. Often at the beach we find ourselves 'shopping' for a certain color or size to fulfill out an abstract idea. We never think of making a story with our work [in contrast to the art work of Tess Felix], aside from the larger story of plastic in our lives."[32]

Plastic is mediumistic. There is a "ghost" in the medium. The trawl of plastics comes unbidden with their narratives, lost and half-lost, to the artists. They come as a new reality, most often as deteriorated and deceased, shaped and refigured on their own, and "ghost" their presence toward another reality of imagination, narrative, and ethics. It is in the individuation that the ghost resides—the ghost in the plastic, the rhythm and fever of the dead and the disused—as magnetic, fascinating, and revelatory. There is an inner life to plastic inside the laboratory; this is another version of an "inner life" hidden from view, growing in surrealistic rhythm and away from roiling humanity. Gabriel Gee notes that "the leftovers made of plastic carry with them a narrative of interconnected history, that somehow bridges the gap between nature and the artifice: the plastic is the archetypal artificial material produced by men's industrial activities, but is ambiguously redeemed in its return to the sea, to the natural environment, where it pursues its own species' evolution."[33] These two lives (between "object made ready" and object as "found") are connected by a tumultuous "outer" middle life of overuse, misuse, and abuse.

The aesthetic energy that runs through these images facilitates the latent energy, the sudden transfiguration of the ready to hand. The dead and discarded are alive again in their inherent plasticity. Lang and Selby Lang's *Bedford Blue* (figure 15) dialectizes the fragility and power of the ocean with the abundance and aberration of beach plastic: a curated project in blue plastic, a dazzling "chromophilic" arrangement in subtle shades of blue. "Once the shelves were mounted and *Chroma Blue* was hung," they write,

we piled everyday objects (in blue) on the sweep of shelf. Along with the thermoplastic junk of our throwaway culture, we tossed a selection of international

---

32. Lang and Selby Lang, "Interview."
33. Gee, "Nature, Plastic, Artifice."

Figure 15. *Bedford Blue* (2011) by Richard Lang and Judith Selby Lang.
Used by permission of the artists.

single-use bottles from Korea, Japan, China and Malaysia to show how the ocean currents are the great conveyors bringing debris from all around the Pacific Rim to us on Kehoe Beach. As brackets for the ensemble we hung our photographs of blue nurdles. Nurdles or pre-production pellets are the result of the fractionating process of forming hydrocarbons into easily shipped bits; in this case they were colored blue before being made into bottles, bins, or bags.[34]

Bedford Blue evinces a nonclassified arrangement; it is about disparate plastic objects teasing each other in a new-found intimacy. The energy in *Bedford Blue* is remarkable for its intimate arrangement in dynamic blue and holds up the objects in their original hue and shape. This practice connects with Joseph Cornell's philosophy of "poetic theatres" where, as Katharine

---

34. Lang and Selby Lang, *The Great Wave*. Citing the importance of colors, material, and visual affect, Tuula Närhinen notes that "arranging the colours, I felt as if I was in possession of the whole spectrum. To see the crayons harmoniously lined up was so satisfactory that to put them in order was at times almost as enjoyable as the fun of drawing. To work with plastic debris made me relapse into my old naive habit of arranging colors" ("True Colours of Twilight," 120–121).

Conley describes, "boxes are joined by chance with salvaged pictures intentionally collected and arranged. They conjure three-dimensional worlds in which fragments of everyday things, from pasted images to dime-store jewels, create a parallel, otherworldly universe imbued with nostalgia. Cornell's 'shadow boxes' summon ghostly shapes and images in the mind's eye of the viewer."[35] They come from the "other-world," as it were, to connect again with the world that we live, consume, and suffer in. They refer to a lost time, the "untimely" that comes to speak about our time. "This juxtaposition," Conlay rightly observes, "confirms the importance of collected, collaged, and assembled things as material, automatic expressions in which assembled objects substitute for words and phrases in order to articulate each poetic artist's most human reality, namely the outward expression of the intensely personal human experience of living, thinking, and being in the world."[36]

Deriving from the French word *coller*, meaning to paste (first used by the poet Guillaume Apollinaire), collage is a "plastic process" executed as much through artworks such as Kurt Schwitters's *Merz* as through the installations and exhibitions of the plastic artists (*Bedford Blue* or *Impressions Plastiques*) discussed here. It is what Daniel Kane calls the "provocative joys of juxtapositions and mysteriousness" without overlooking what constitutes "authority, identity, voice, originality, sincerity and art." This is an "intuitive self-definition of the artist among objects."[37] Plastic collage adds to the rich and complicated tradition of collage that prospered through the twentieth century. The distinctiveness of such an artistic representation lies in collaging with diverse forms, shapes, and sizes of one material. The material variety brought into the canvas by Cornell or the page by William Burroughs or Frank O'Hara is replaced by mono-material aesthetic execution and obsession.[38] *Bedford Blue* has a subtlety in the arrangement, invested in ellipses, untold stories, and tangential thinking. It figures what Rosalind Krauss calls the "constant superimposition of the grounds" that "inaugurates a play of differences which is both about and sustained by an absent origin."[39] This art installation builds an exterior experience in plastic heterogeneity but constructs

35. Conley, "Collecting Ghostly Things," 268.
36. Conley, "Collecting Ghostly Things," 279.
37. Malevich, "Spatial Cubism," 60.
38. See Cran, *Collage in Twentieth-Century Art, Literature, and Culture.*
39. Krauss, "In the Name of Picasso," 20.

an interiority that connects the artists with the world, our fraught and febrile nexus with plastic.

Most "plastic images" are auratic, living through an instant of aesthetic reconstructionism. It is an embodied aesthesis—a kind of "experience of envelopment"[40]—coming from the corporeal, visual, sensorial, and imaginative experiences. Conley observes that "having been found, collected, turned away from its original function, and displayed by a surrealist, the object represses its 'manifest life'; its transformation generates a veritable force field (*champs de force*), whereby what was formerly manifest becomes latent, revealing ghostly energies inherent in the object's former manifest life."[41] Plastic materials washed up and tossed about have voices—materially intoned and scripted—that connect chiasmatically with the voices of the artists. This is the point of a physical, figural, conceptual, and aesthetic intertwining. I see in this negotiation both a connection and separation—a syncopic formulation where contact with plastic forms its own revulsion and retraction. Seeing and feeling plastic are experiencing and discoursing about the anguish that the "degenerate" plastics produce. Here is a joy in art which comes close on the heels of an agony.

"To know materials," Tim Ingold points out, "we follow them."[42] Lang and Selby Lang, among others, "follow" the material to build what Israel Scheffler calls a "characteristic tension"[43] between the artist and the material. The material as medium—"numerous, complex, visible, weighty" (in the words of Henri Focillon)—generates possibilities.[44] The material invites its own transformations through a distinct, controlled interaction; it comes in abundance, in multitudinous forms and incarnations, into a space of aestheticized composure and poise. It's like touching the plastic twice: once in use and the next in disuse. If the use is design art, the disuse is design art with a difference. The turn to plastic is thus multiply affective and designative. Affective experience in plastic arts comes from a sense of an experiment and trial: experience as "related to the Latin *periri*, and *peritus*, which refer to making a trial, a trial which possibly may not turn out fortunately. 'Peril'

---

40. Conley, *Surrealist Ghostliness*, 12.
41. Conley, *Surrealist Ghostliness*, 18.
42. Ingold, "Toward an Ecology of Materials," 435.
43. Scheffler, *Philosophy and Education*, 110.
44. Focillon, *The Life of Forms in Art*, 96.

is derived from the same source. Here experience is experiment, a questioning trial."[45] The physicality involved in the collection—orientation, ordering, and patterning the plastic materials and their eventual aestheticization—speaks of a "fluidity" of psycho-corporeal manifestations that are strongly intertwined. The coming together of the material revises the signifying potential of plastic matter as one system of signifiers collages and conglomerates to produce a different meaning with and out of the matter. Närhinen and Lang and Selby Lang construct such affective assemblages. The aesthetic transformation that plastics through combinatory poetics bring tempts us to believe that plastics preserve and contain in themselves "art" or art-expression. The ignored and chanced upon hold a difference to itself in that the "spooky" *Love Hand* and the *Dinosaur* and the *Lamb* declare how intensities and aesthetic emotions have stayed "blocked" in the material, awaiting "events of experience." The art-matter vibrates with a supplement that does not necessarily have to be a nomothetic signifier. For instance, Lang and Selby Lang's *Known Quantity–Combs* (figure 16) accentuates the "presence" factor of the material plastic.

The representational appropriateness of art as discarded combs can slip into "suspension," enabling the plastic to "unblock" what humanly was never envisaged and experienced. The images are unsettling and make possible a "presence," a fluidity in experience, and a trial that continue to marvel us. The collaging and chroma, the wear and the tear, of the combs speak of an aesthetic intensity, generating an affective supplement. We encounter a reciprocity that works at three levels of aesthetic affect: the things in the world and their fate within the human world, the connective dynamics between the plastic material and the artist's imaginative breadth and insight, and, finally, the material as it exists and how it has been made to exist through a separate artistic frame. This is the "murmur" of plastic as it sounds the different layers of material-aesthetic transaction. There is an aesthetic rhythm in art that declares a "plastic time": the action of time beginning with the production of combs, their disuse, the fragmentation and weathering, washed-up time, the collection, the preservation, and finally the production of art. But this diachronicity cannot ignore the "harmonic time" involved in the construction of art. Plastic arts have a resonant or intrinsic time that comes

45. Ballard, "The Nature of the Object as Experienced," 114.

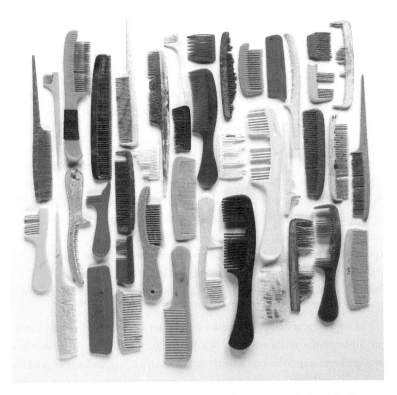

Figure 16. *Known Quantity–Combs* (2011) by Richard Lang and Judith Selby Lang.
Used by permission of the artists.

from its organization, contemplation, and aesthetic unity: "The time of the work radiates, so to speak, around the prerogative moment represented. The latter makes a structural center from which the mind moves backward to the past and forward to the future in a more and more vague fashion until the moment when the image fades gradually into space."[46] The prerogative moment of creation leaves the history of the combs obfuscated and on the edges of speculative haunting. Time trembles; the moment glows through a plastic rhythm; art *presences*.[47]

Breton proposes doubt on reason and rational thinking and plastic art similarly suspends the conventional "logic of art" to build "objective chance."

---

46. Souriau, "Time in the Plastic Arts."
47. For more on "presences" see Ghosh and Kleinberg, *Presence*.

The plastic artists cut into the logic of artistic representation through a material and a mode of aesthetic reason and value. They do not have the choice in supply but have the choice in choosing—a fresh dialectic in subject-object as distinguished from the initial dialectic when the object was a "toy" in the hands of the subject. The plastic turn lives within a desire that is resourceful and inexhaustible, revolting and revolutionary, conscious and subliminal. Strewn across the sea-shore, left awash on the shorelines, regurgitated in forms numberless, the "distant" plastic comes near in a transformative art in which suddenly the proximate becomes unfamiliar and the abandoned becomes our possession. Material-aesthetically, the plastic arts become reclamatory, through both material and affect—a remembrance of things we think are lost. Interestingly, the lost objects stay unknown in their value and aesthetic power. All these art works demonstrate that plastic objects are not identifiable with a name because through ceaseless weathering the "origin" has been diverted into an Other of unclassified identity. These largely unnamed plastics find their place with each other to "image" an object that they are not and yet cannot be separated from (for instance, the *Dinosaur* and the *Lamb*). The fragmented existence in such poetics of sharing builds towards a greater picture for which they are responsible and to which they are made responsible.

Desire is central to the material-aesthetic phenomenon. Närhinen's *Impressions Plastiques*, *Soleil Levant*, *Soleil Couchant*, and *Frutti di Mare*, Lang and Selby Lang's *Bedford Blue*, and Evelyn Rydz's *Floating Artifacts* open themselves to the "possibility of finding." Daniel Barbiero explains this state as representing "a way of being in the world that sees in the world certain possibilities as they are illuminated or disclosed through our desires. *Disponibilité* just is the availability or openness to the unexpected, a susceptibility to noticing the strange coincidence and chance meeting that turns out to be of unanticipated significance."[48] The artwork embeds in "availability" a set of things that has desire, chance, and intention and is undefined. Innumerable plastics wash up onto the artistic consciousness, their availability

---

48. Barbiero, "The Object as Catalyst." See also Breton, *L'amour fou*, and Cardinal, "André Breton." "The surrealist's characteristic expectancy and *disponibilité* are thus essentially the same whether he is about to encounter a new image or whether he is about to meet the woman with whom he may well fall in love. It can be readily appreciated that there is an analogy of mood in the attitudes of the surrealist poet and the lover: both are prepared for unexpected pleasures, both are willing to abandon reason" (Cardinal, "André Breton," 116).

underwritten by a desire. And all plastic projects are events in desire, in spaces where occupancy is about finding a relationship among objects, material reality, history, and a life that the object forces us to think outside itself. This is about desiring the object and the object's ways of making itself *available*. All the art-works spoken of thus far, speak of a "creative act as an occupation." Rick Dolphijn explains that

> Deleuze uses "to occupy" thus not primarily as a socio-political term but as central to *a creative act*. It signals how art itself, rather than the artist/activist, *as an event*, searches to give form anew. To occupy, then, does not start as a political act or as social engagement, it concerns the event that reveals another politics, another ecology and another sociology always already there. "To occupy" conceptualizes a force which sets about creating or *materializing* a new environment in which objectivities and subjectivities take shape. To occupy is then not an activity which has to start with human activity. On the contrary: *it is creativity itself* and its ability to involve others (and the way these others, like us, allow themselves to be involved with it), that mobilizes the true revolution.[49]

The turn to plastic arts builds its own "occupancy" as the "washed down" material deterritorializes the familiar and the obvious; art with "altplastics" (my representative term for all kinds of plastic discards) unfixes the dynamics of "to occupy" and builds its own relational poetics with matter, intention, design, emotion, and affect.

The plastic objects washed ashore and combed out of the beaches—existing as art objects by themselves and as objects for the artists—contribute to the "surrationalisme" whereby the physical world makes itself available in innovative and provocative ways. This is the "crisis of the object." Rudosky argues that

> instead of the object or the material world being second to thought as he relates to "reductive" thinking, the object will now come first by way of "inductive" thinking. Thought, or conscious perception then is relegated to the object and the material world. Breton also suggests that instead of being underestimated in thought, the object will finally be "beyond" thought. We can also see this as being "beyond" reason, or beyond comprehension in logic. As such, Breton

---

49. Dolphijn, "The Revelation of a World That Was Always Already There," 192.

rephrases this difference as a kind of movement and suggests that thought will never actually attain the "object" as it will recreate itself at a vanishing point, "à perte de vue au délà" (beyond sight). If the "object" in Breton's surrealism is a kind of external material to the individual as well as part of the physical world, and we remember that Breton's quest is to go beyond "le monde soi-disant cartesien qui [nous] entoure" towards a world which reduces antinomies and the separation between the unconscious and consciousness, it becomes understandable that Breton wanted to break the barriers between the external, physical "objective" world, also related to consciousness, and the interior psychic interior world, related to the unconscious. Breton does not transgress the barrier per se, but rather puts it into a perpetual mode of relation.[50]

Plastic-arts have achieved such a mode of dialectical relation whereby the objects lying across the planet have subjectivized their presence and destabilized rational thinking around art objects. The turn to plastic has magnified and multiplied the aesthetic possibilities with objects: art-material has a new breadth to choose from and offer. Found art, art as collection, artisanal collage and orientation combine to produce the plastic affect-ability. The line that separates plastic-arts from plastic-infested life is at best indistinct. Through art, life becomes more inclusive and articulate than art. Have plastics in the sea and its beaches become our newfound commonplace? The plastic turn has restituted and restored the abandoned and the forgotten, giving a new meaning to the commonplace—the defamiliarization mystique—whether in the form of toothbrushes or bottles or combs or hair pins.

Nicolas Bourriaud's "relational aesthetics" is predominantly about a set of relations conceived of as social relations. Less concerned about objects, Bourriaud accentuates the function of interstice and precarious art as political strategies to produce counternarratives against global capitalism and other forms of trading and productionist mechanisms. Bourriaud maintains that when capitalism makes us "live in a finite, immovable and definitive political framework," precarious relational art "present[s] the reverse postulate: the world in which we live is a pure construct, a mise-en-scène, a montage, a composition, a story and it is the function of art to analyse and re-narrate it."[51] The plastic art-works discussed here have their own *débordement* in that they exceed their simple frameworking

---

50. Rudosky, "Breton the Collector," 45.
51. Quoted Potgieter, "Critique of Relational Aesthetics," 2.

within relational aesthetics. Plastic affect can be disarming to a point where art recreates a reality through a suspension of disbelief. This is relational aesthetics with a difference. Art is not exclusive as plastics project intersubjective relations between human-nonhuman situatedness and their sociocultural contexts. Working beyond Bourriaud's emphasis on inter-human relations, we can argue for a greater network that relates the human, nonhuman, the nonliving, and the earth. The plastics coming together in a variety of relations are the performative that brings out a busy space and form: it is unstable through multiple narratives invested in such relationalities. This instability is the plasticity of a system of heterogeneously abled and multitudinously potent plastic-particles. I call this an "eco-logical interstice" where exchanges across consciousness and systems of thought (eco-cultural-historical and also commercial and sensorial) come to "construct" plastic art. The unusable and inexchangeable plastics, in a relational aesthetics, produce a plastic event. What the plastic pellets and objects do is that they come with some kind of conceptual opacity that neuters all instrumental dimensions with which we are habituated to enjoy and live plastic. This, in a Kantian way, encourages the mental faculties to build their own relational aesthetic wholeness without ignoring some conceptual determinacy. It is the frustration with plastics, their formlessness and aberrational state of existence, that inspires the plastic imagination, prompting nonformulaic patterns of thinking. I see Bourriaud's emphasis on the "infinite tendency" of a work of art relevant: plastic materials have no a priori "useful" function but come together conceptually, sensibly, and sensorially to exceed conventional frameworks of understanding. Relationality declares a tie-up with chance, randomness, and the "unconscious" in art. Plastic produces its own aesthetopia.

Plastics, commercially produced as well as those that are beach-combed (the differentially produced), invisibilize the "inbetweenness" that exist between these two states. It is as if we are holding the two ends of a rope without knowing what happens and what the state of the inbetweenness is. It is post-processual both in the sense of the post-ness of the processed plastic that we use and reject and the processuality that plastics undergo in their state of human rejection. Gilbert Simondon's ontogenesis explains this plastic evolution well. Plastic's differential growth ("individuation") provides surprising and uncanny identities as they make art *possible* through a "rhythm of becoming."[52] Lang and Selby

---

52. Combes, *Gilbert Simondon*, 22–23.

Lang's paired objects, *Dinosaur* and the *Lamb*, is a substantial testimony to such ontogenetic formations. The material loses its equilibrium in an inchoate state that makes for the affective aesthetic dimensions. Plastics dephase, build their own discontinuities with varying milieu and emergent complexities, and rebuild their aesthetic resonance. Plastics demonstrate matter-flow, and their peripatetic ways urge artists to intuit their collection and aestheticize the consequent assembling and construction. Tim Ingold notes that "instead of the concatenation of the *chaine operatoire*, where both techniques and forms go from point to point, we have here an unbroken, contrapuntal coupling of a gestural dance with a modulation of the material."[53] Plastics modulate to surprise the consumer and again, modulate to surprise the artist and the viewer. The artist's turn to plastic involves encountering "meshwork" (as Ingold calls it).[54] Out of the laboratory and the industry and on the sea beaches, plastics modulate form and expression, takes on surprises and surprises themselves, write two dominant histories where one is about the material-scientific and consumerist success and the other a helpless capitulation to a recalcitrant degradation. This gives birth to the "plastic subject" that plasticizes its emergence, usuability, denouement, nonutility, and materialization through an undulating trajectory of manifold becomings, ecological materiality, and eco-cultural occurrence. We follow plastic as much as plastic follows us.

Plastic, as "found," is art and no-art; it has a design and aesthetic to it and also submits to artistic "reframing" through a different order and affect. The turn to such a kind of plastic is dialogical, and this extends to greater dialogism through art in communication with the viewers—the ambit of communication keeps revising its frameworks of interaction and intersubjectivity. What interests me is how plastics change place or site and change their space and sitings with it: from laboratory to the cultural-economic world to the garbage bin to being discarded on land and at sea and then to traveling

---

53. Ingold, "Toward an Ecology of Materials," 427–442.

54. Ingold writes, "In a sentient world, by contrast, things open up to the perceiver even as perceivers open up to them, becoming mutually entangled in that skein of movement and affect which Merleau-Ponty famously called 'the flesh,' but which I have characterized, more accurately I think, as the meshwork. In the meshwork, the 'flesh' of phenomenology is unified with the 'web of life' of ecology. Thanks to their entanglement in the meshwork, my seeing things is the way things see through me, my hearing them is the way they hear through me, my feeling them is the way they feel through me. By way of perception, the world 'coils over' upon itself: The sensible becomes sentient, and vice versa" ("Toward an Ecology of Materials," 437).

unknown places before reemerging to find their place with these artists and change the spatiality of their presence and impact. There is an uncertainty and indeterminacy in their spatial and locational turnings and returnings. Arnold Berleant's "contextual aesthetics" has certain features: "acceptance" as "an openness to experience" while immediate, selective, and restrictive judgments are suspended; next are "perception" and "sensuousness"; the fourth dimension is "discovery" as "a freshness and sense of new possibilities"; then come "uniqueness" and "reciprocity"; next is "continuity" where the "distinctions we draw from a reflective distance between the constituent elements of the aesthetic field" disappear gradually, and "the divisions and separations that we impose on experience to help us grasp and control it melt into continuities. This is the primary milieu of aesthetic experience and secures its contextual character"; finally, comes "engagement" where boundaries fade away and we become vulnerable and "multiplicity" that expose us to the "sites for aesthetic involvement, limited only by our willingness to participate and our perceptual sensitivity."[55] In line with Berleant's idea of "continuity" and the "aesthetics of engagement" plastic art does not stay confined to the domains of traditional aesthetic appreciation; it enfolds a wide range of emotion related to human and nonhuman experiences. Here lies the interested contextuality to plastic-appreciation. Altplastics exceed human imagination and art-abilities, and when collaged and reconstructed, they continue to self-exceed themselves through assimilative frameworks of understanding, which are local, topical, political, and eco-cultural. This turn to plastic is poly-sensorial, depending on perception and experiences that make artists and viewers confront the "intractability" of plastic as a material in its multiplicity and dispersion. In art, plastic is "dematerialized" to construct a continuity with the artist, the viewer, and the art material. In the material-aesthetic engagement, plastic, most often, exceeds its objectivity and, in the consequent defamiliarization, the "aesthetic listening" starts to vary. Plastics in "free supply" are not simply material commodities attached to an economic bandwidth (the Marxist use-value). Bestrewn across the beach and unbeknownst to the artist and her subjective desires and intentions, these objects, upon discovery, build a connection to subjectivity. They hold a meaning but, in turn, make meaning available to the discov-

---

55. Berleant, "Ideas for a Social Aesthetic," 26–29; Clark, "Contemporary Art and Environmental Aesthetics," 359.

erer, the collector, and the artist; they come together as a "system of objects" in art with their own narration and artistic operation.[56] This system of objects comes with a newfound aesthetic emotion and a strong subject-object rapprochement wherein meaning is prismatic and world-directed. Plastics, when in use, hold a world for us; when out of use, they hold a world for us too—a microcosm bearing snapshots of our material world, the material culture and material depredation.

The artists, under the stress and spread of plastic, cannot remain with a hard-edged "individualized consciousness" that ignores their experiential connection with the lived world and its sociocultural persuasions. Invested in what Suzi Gablik calls "connective aesthetics,"[57] plastic artists refuse the "monocentric mythology of the artist." Plastics demand a connective radicality that requires, as Gablik argues, "a consciousness that is different from the structural isolation and self-referentiality of individualism. In the post-Cartesian, ecological world view that is now emerging, the self is no longer isolated and self-contained but relational and interdependent."[58] Närhinen, talking about her *The Mermaid's Necklace*, notes:

> I was shocked to find several square meters of shoreline awash with agglomerations of small plastic pellets. I must have passed the same spot hundreds of times without noticing anything out of the ordinary. But in order to perceive, you need to look for, and to look for you have to know that it exists. To actually see the plastic pellets and to tell them from similar grains of sand such as the translucent crumbs of quartz, one has to be familiar with the concept of mermaid's tears. I developed a sieving method for separating microplastics from the rest of the flotsam. The installation shows the sieves along with the final result, a necklace put together of tiny pieces of plastic. *The Mermaid's Necklace* looks beautiful but the story behind it is sad.[59]

---

56. For an elaborative note on the spoken system of objects, technological plane, and everyday structurality, see Baudrillard, *Systems of Objects*.

57. Gablik, "Connective Aesthetics." Gablik writes: "Connective aesthetics sees that human nature is deeply embedded in the world. It makes art into a model for connectedness and healing by opening up being to its full dimensionality—not just the disembodied eye. Social context becomes a continuum for interaction, for a process of relating and weaving together, creating a flow in which there is no spectatorial distance, no antagonistic imperative, but rather the reciprocity we find at play in an ecosystem" ("Connective Aesthetics," 86).

58. Suzi Gablik, "Connective Aesthetics," *American Art*, 6, no. 2 (1992): 2–7.

59. Närhinen, "Between the City and the Deep Sea," 158.

This is also an "aesthetic arrest" for Lang and Selby Lang,[60] a cessation of "habitual thinking" and an opening to a possibility that is "necessary in the face of daunting facts" that are political, social, and cultural. Unhinging habits of aesthetic sensibilities, the material-problematic inspires a separate vein of "aesthetic listening" in the doing and performance of art. The material-aesthetic submits to an entangled ego-subject whose credibility is in a participatory ethics of "living together": human, the nonhuman and the material. This is a deeply interactive mattering, an actualization through "being-in-the-world."

---

60. Lang and Selby Lang, "Interview."

Chapter 4

Plastic Literature

Where is plastic? Here, there, anywhere, everywhere: surely, somewhere. From plastic bottles to plastic money we are (en)plasticized: plastic trees, plastic birds, plastic soil, plastic water, plastic stomach, and a plastic planet. This is our "plastic contract" beyond annulment and revocation, submitted to renewal every moment, every breath-time. We are living inside a plastic bomb; the sound of its explosion cannot be heard but is experienced through the flow of time. This generates a waste-apocalypse, an irremediable fate of extinction through a gradual but certain course. Plastic is an explosive but its effects return to us only after we have long forgotten its purpose. Used plastic leaves behind a memory of guilt; plastic waste is witness to a grind of biotic extinction. By keeping it out of sight, by looking away, we allow it to ensnare us, to suffocate the planet.

Folded Waters

Through plastic, the sea has become far more striated than Deleuze would admit or Foucault could fathom.[1] The "more than human" assemblages have taken the land-sea stories beyond mere fish, ships, waves, commerce, and voyages. We are getting wet in a different way, and the "wet ontology" of sea-liquidity is far from being static. Plastic is our new "hydromateriality."[2] Ever since plastic marine pollution was first discovered in 1972 in the North Atlantic by Edward J. Carpenter and Kenneth L. Smith, our conscious and ignorant cohabitation with a new-found-sea has begun. Initially balked and bandied about under the governmentality of plastic industries that multiplied its production and reach through the logic of user-friendliness and "planned obsolescence," research into the relationship between plastic and the ocean took some time to highlight the menace and doom until Captain Moore encountered the North Pacific Subtropical Gyre in 1997—"the unspooling of this man-made horror show in the Pacific, and possibly in all the earth's ocean."[3] Sea-plastic, in fact, has become a much bigger problem than when oceanographer Curtis Ebbesmeyer coined the term "garbage patch." Projections suggest that by 2050 plastics will be more prevalent in oceans than fish.[4] Either suspended among planktons or caught as surface soup, in gyres or as submersibles in deep sea and on ocean floor, plastic has built its own blue spread. Billions of fish rise to the surface to feed, and they are, as Marcus Eriksen writes, "ingesting nonnutritive plastic, which gives a false sense of satiation, creates potential for intestinal blockages, and introduces a buoyant material in their bodies that forces them to expel energy they can't spare to get back down deep before the sun rises. Like that piece of foamed polyurethane from a flip-flop with chewed edges: a fragment of that in the gut of a one-ounce deep-sea fish is like you or me swallowing an empty two-liter bottle and trying to swim to the bottom of a

---

1. The title for this section is borrowed from artist Rydz's "Folded Waters."
2. See Anderson, "Relational Places."
3. Moore, with Phillips, *Plastic Ocean*, 92.
4. See Andrady, "Microplastics in the Marine Environment"; Auta, Emenike, and Fauziah, "Distribution and Importance of Microplastics in the Marine Environment"; Avery-Gomm, Borrelle, and Provencher, "Linking Plastic Ingestion Research with Marine Wildlife Conservation."

pool."[5] Marine plastics cannot exist in a "commonplace way"; they refuse to be dispersed, incinerated, mineralized, and mined as one might do with many other material wastes. There is an inextricability that attaches to plastic where beyond a point the environment and the plastic are inseparable, become one.

Does the plastic sea leave us with a sense of self-exclusionary consciousness of discard, filth and garbage? How does discardism build its own intricate ways of attachment? Plastics have come with a new idiom of re-surfacing, a return "from the molar category of garbage to the order of the particular and molecular" producing "material and affective entanglements."[6] This makes our detachment an illusion and the consciousness of "harm zone" unsustainable. The plastic as litany and as garbage builds a molecularity of bashed borders of hyper-relationality where the consciousness of plastic is hauntological.

Nature's metabolism and levels of assimilation are under serious pressure— the tragedy of the commons[7]—and most conventional assumptions and inferences about plastic dissemination and travel are proving alarmingly wrong. It is true that thinking about nature begins when our living amidst nature is in crisis. Within the framework of what Patricia Yaeger calls "Ecocriticism$," the "ocean as *oikos* or home rolls under, beneath, and inside the edicts of state and free market capitalism. We've left the possibility of wilderness or pastoral for the roller coaster of capital." The plastic sea is the "fleshy entanglement of sea creatures, sea trash, and machines."[8] It demonstrates the rapacious profit-making capitalization underpinned by an ethics of irresponsibility where the "sea is just another site where human relations take shape and connect through low-cost hardware and the freedom of an unregulated environment."[9] It is more "techno than ocean." Although this chapter is not a scientific paper intended to detail minute research on plastic in the sea, yet we ought to shudder at the realities that 80 percent of ocean pollution begins on land, that "almost seventy five percent of the world's fish stocks are already fished up to or

---

5. Eriksen, *Junk Raft*, 131.

6. Huang, "Ecologies of Entanglement," 104.

7. Hardin, "From 'The Tragedy of the Commons.'"

8. Yaeger, "Sea Trash, Dark Pools, and the Tragedy of the Commons," 530.

9. Yaeger, "Sea Trash," 533. Also of interest is Cranstoun, "Ceasing to Run Underground." It is the technologization of the deep sea through cables, winding machines and other sounding devices. See also Rozwadowski, "Technology and Ocean-scape: Defining the Deep Sea in Mid-nineteenth Century." On network archaeology, see Starosielski, *The Undersea Network*.

beyond their sustainability limit," that five out of six plastic bottles are not re-cycled and that a plastic bottle can take four hundred fifty years to degrade, that the ratio of plastic to zooplankton is 36:1, that 61 percent of the plastic in the ocean is less than 1 millimeter in size, that each year plastic litter in the ocean kills hundreds of thousands turtles, dolphins, whales, and other marine mammals and one million sea birds.[10]

Thinking the plastic sea is, as Audra Mitchell writes, "thinking without the circle": no matter how and to what length we stretch and extend the cir-cle of existence and however much we think that stretching can "circle out" the harm, the intransigent reality of being "always-already encompassed and interpenetrated by marine plastic" remains.[11] Marine plastic has decimated the enlightenment ontology of "hyper-separation" of the human as holding an autonomous space distinguished from every other nonhuman interference. The circle that humans thought created a boundary between themselves and any harm has been permanently undermined. The hyper-relationality of plastic toxicity, an ambivalent entanglement and plastic's "ability to cohere and assert 'force-fullness' on massive (and micro-) scales"[12] have transformed our definitions of existence, motives, and desires in the plastisphere. I call this "plastic imperatives" or cosmopolitan motilities.

This is our modern day "plastic contract": our shared plastic habitat where all of us—human and nonhuman—are plastic creatures, a kind of "inter-implications of forms of life with inorganic forms."[13] In fact, the community of plastian[14] discovers a new meaning in "entanglements"—an entangled event happening through interspecies connection and what we call plastic de-bris in the form of "fishing line slowly strangling seals, and plastic bottle rings giving hourglass figures to turtles. These are physical relationships—contact between animal bodies and plastic materials."[15] Catherine Phillips sees this state of the plastic sea in terms of "traces"—of vestiges, "enduring fragments," understood as "out of place" and abject with a "half identity [that] still clings to them." This phenomenon is hugely entangled with prospects of unexpected history and narrative of survival and sustenance. Within discard studies, such

---

10. See Rothschild, *Plastiki*.
11. Mitchell, "Thinking without the 'Circle,'" 77.
12. Mitchell, "Thinking without the 'Circle,'" 78. See Shaw and Meehan, "Force-Full."
13. Grosz, "Habit Today," 235.
14. Intemann and Patschorke, *Plastian the Little Fish*.
15. Wolff, "Plastic Naturecultures," 26.

trace and micro-waste have complicated the issues of biotic survival. When the traces are ignored and untraceable the traces promise futuristic instabilities.[16] Kim de Woolf is right to observe that "in the plastisphere, humans, disciplines and ocean creatures 'become with' plastic. As they travel, indeterminate plastic species bodies not only 'become-together,' but 'become-apart,' in practices that sort kinds of materials and kinds of species, both living and nonliving. And plastics are named as potential species in return."[17] This process involves a geontological power that makes claims on a new mode of governance, one that works on the indistinction among life, death, and nonlife (within the frame of my argument here it is the plastic). Elizabeth Povinelli's arguments about "potentiality" and actuality enable me to think of the "capacity" of plastic to change life—the nonlife and life-losing distinctions, staying inter-territorialized, and, hence, reconfiguring the agency of governance and governmentality.[18] If the plastic sea has triggered new modes of power to measure and control and think of ways to informationalize and figure the materiality of the ocean, this governance is not dependent on one side of the agency; the sea itself is agentic, with plastic making the modes of discursive governance ambivalent through nonlife potentiality. Jessica Lehman sees the "World Ocean" as the fourth figure (moving beyond the three figures—the Desert, the Animist, the Terrorist—that Povinelli argues for[19]) of geontopolitics. She argues that "as a possible fourth figure, the World Ocean is, like the other figures, another complex entity that occupies the institutions of contemporary planet-scale government."[20] The *emerging* sea affects the understanding of power and geopolitical monitoring and oceanographic mensuration with nonlinear dynamics, disequilibria, dispersive materiality and an excess. This excess of the nonliving—*potenza* of plastic—is what blurs the distinction between life and nonlife and, thus, geontologically, provides a fresh sea-sense.

Is the sea Emily Dickinson's "deep Eternity"? Is it being seen as the "birth of the aquarium"? Are we still living the sea in our minds, outside our culture and in a separate nature, as hypernature? The reality is different. The

---

16. Phillips, "Discerning Ocean Plastics."
17. De Wolff, "Plastic Naturecultures," 27.
18. Povinelli, *Geontologies*; see chaps. 2 and 7 for some interesting observations.
19. See Povinelli, "Three Figures of Geontology."
20. Lehman, "A Sea of Potential," 121.

hydro-materialities of the plastic sea reterritorialize seascapes and thalassic consciousness in meshed eco-cultural-economic-political ways. The fluidity of the sea—its modelization, modern mythification, and the relationalities—has changed. Cecilia Chen, Janine MacLeod, and Astrida Neimanis explain that "with its etymological root in the Greek *oikos*, meaning 'home' or 'dwelling,' the prefix 'eco' indicates *where we live*";[21] in fact, eco is where we belong with all the expansiveness, inclusiveness, and assimilativeness. With the ever-rising plastic, the eco is "unhomely," where home-coming is a process in anxiety, threat, and death: all changing, all vanishing gradually, everything together unfixed and ungridded. The history of the sea has changed: caught within its extra-national and heterotropic modes of becoming, we are irrevocably in the midst of the plastic sea.

Plastic Literature

Can marine plastic behavior and agency—its planetarity and immanence—be the material-aesthetic to critique *how* we think of literature in a globalized world? Are we not in the midst of plastic poetics, the performatics of which I call "plastic literature"?

Stacy Alaimo argues that "climate change, sustainability, and antitoxin movements make environmentalism a practice that entails grappling with how one's own bodily existence is ontologically entangled with the well-being of both local and quite distant places, peoples, animals, and ecosystems. Campaigns against plastic link not only coastal regions but also inland zones to the mushrooming plastic found in the oceans."[22] Across the land and sea, plastics have declared their presence in forms that are durable, malleable, degradable, nondegradable, sinister, and toxic. In fact, plastic "not only spreads while maintaining its molecular form, but the plasticizers that are added to plastic (one or more of a possible 80,000 chemicals added to make plastic pliable or pink or heat-resistant) leach and off-gas; detached from the polymer bond, they are able to move into the surrounding environment and whatever bodies may be found there."[23] And, again, as "plastics

---

21. Chen, MacLeod, and Neimanis, "Introduction," 14.
22. Alaimo, *Exposed*, 131.
23. Davis, "Life and Death in the Anthropocene," 350.

gain in toxicity their value depletes, they are cast off, re-entering market chains for what little profit can be made from recycling, spreading their accumulated toxins wherever they go. They are then sifted, filtered through, recognized for their worth by those who cannot afford to participate in this throw-away culture, for those who are also placed elsewhere, out of sight of the markets of capital that rely on invisible labor in order to perpetuate this system."[24] Within an exploitative, garbological, and fast-capitalist system, plastic continues to spread and be recycled, enhancing the complex networks of access and excess. The translocationality of plastic through plastic pervasiveness and scopious contamination represents a spatial consciousness wherein plastic seeps into salt, water, soil, animal and human bodies, media, and our psyches, mostly in nonlinear, nonhierarchical, fragmentary mobility and fluidity. It does not discriminate between its point of origin and eventual destination, race, ethnicities, color, religious affiliation, cultural heritage, or political leanings. Plastic, thus, provokes connections/comparisons with a difference that speaks of becomings, dispersions, and immanence.

Max Liboiron argues that plastics "move knowledge away from concrete, understood, and well-measured matters of fact towards the lesser-known and less-bounded elusive systems of influence that challenge not only practices of pollution control that have worked (at least in theory) for a century, but also broader theories of what pollution is and how it works."[25] It is interesting and surprising to find plastic materials sinking into the sea floor at all depths from "inter-tidal to abyssal environments."[26] But it defies precise explanation how marine debris *uncannily* reaches or purveys such depths. Plastic encroachments can hit and poach at any levels and can also lead to the formation of *wrack*, which Murray R. Gregory defines as conglomerates of "natural flotsam, of both marine and terrestrial origin (seaweeds and plants) together with jetsam of *indeterminate* sources . . . [that] are often ephemeral, dynamic and seasonal environments and also tend to accumulate significant quantities of manufactured materials, in particular those made of plastic and other non-destructibles."[27]

24. Davis, "Life and Death in the Anthropocene," 351.
25. Liboiron, "Plastics in the Wild," 67.
26. Gregory, "Environmental Implications of Plastic Debris in Marine Settings."
27. Gregory, "Environmental Implications of Plastic Debris in Marine Settings."

Gregory also points out how plastic jetsam is creating new networks of transmission and wayfaring:

> Through the distant past to modern times, these materials [natural flotsam] have . . . attracted a diverse biota of sessile and motile marine organisms— *freedom travelers* (hitch-hikers and hangers-on if one likes). This process has been a mechanism in the slow trans-oceanic dispersal of marine and some terrestrial organisms. . . . The hard surfaces of pelagic plastics provide an attractive and alternative substrate for a number of *opportunistic colonizers.* With the quantities of these synthetic and non-biodegradable materials in marine debris increasing manifold over the last five decades, dispersal will be accelerated and prospects for invasions by *alien* and possibly aggressive *invasive* species could be enhanced.[28]

This opportunistic colonization, invasiveness, indeterminism, graft, and gradient-flow narrate, on a corresponding scale, how transcultural and transnational dimensions of literary thinking work and perform. Plastic, like trans-literary thinking, is "athletic": "it scoots, flies, and swims. It travels without passport, crosses borders, and goes where it is, literally, an illegal alien."[29] It has also been argued that terrestrial ecosystems, particularly those of remote or isolated small islands, could be "endangered by vertebrate species (e.g., rodents, mustellids, cats, etc.) by rafting on matted plastic megalitter."[30] This marine performance—the illegal *migrancy* and stealth *drift*—is a clear pointer to the principle of "trans-habit,"[31] where aggregation, planetarity, co-constitution form its significant performative motors. The transoceanic dispersal and cross-border movements correspond material-aesthetically with what I argue to be the trans(in)fusionist literary,[32] wherein "travel"—horograhic, nonchronometric, telic, and nomadic—becomes a vexed metaphor in transcultural interventions. World-comparative literature is plastic; the material behavior explains the aesthetics and politics of contemporary world-literature formations.

---

28. Gregory, "Environmental Implications of Plastic Debris" (emphasis added).
29. Moore and Phillips, *Plastic Ocean*, 43.
30. Andrady, "Microplastics in the Marine Environment," 388.
31. See my *Transcultural Poetics.*
32. See Ghosh and Miller, *Thinking Literature across Continents*; Ghosh, *Trans(in)fusion.*

In "Sāhitya Shristi" ("The Creation of Literature"), Rabindranath Tagore talks about formation, the coming into being, of ideas and thoughts around a subject—a slow but sure process of literary formation. This is, as Tagore qualifies, a ceaseless process. Fruits coming to the branches do not stay quiet: the hog plum ripens, fills up in juices, puts on color, grows aromatic, gets hard in its interior (the stone), and exceeds the tree of which it is a part.[33] Thoughts follow such a growth: it forms, and its formation exceeds the source of its forming. If an object or a thought settles into a site, building a context of its own, it speaks about itself and also the surrounding, the atmosphere, that contributes to its forming and existence. Identity then is not specific to a thought only; a thought has an identity that plasticizes and relates to other identities contributing to the context that it holds as its own. Togetherness is enmeshed, entangled, and entropic. It is here that a plasticity of literary growth can be located—plasticity in the sense of malleability, flow, and diffusion. Tagore argues that we fail to express ourselves through a brokenness of thought formation and expression; we falter, lack clarity, and miss expressing what we desire to communicate. This insufficiency creates further plasticities of imagination, form, language, and emotion. He considers the "local" as holding on to the demotic rhythm that builds a thought in particles—dispersed and spread out—before they "gyre" into forming a narrative, whether it be the Arthurian legends or the *Ramayana-Mahabharata*. It speaks of the worlding of the local.

Tagore notes that anything that is stable and is considered as not having any possibility of alteration is difficult to believe and accept as existing. The Indian subcontinent has been the seat of literary and cultural mingling, a blender, as it were: discourses primarily from the Hindu tradition have accommodated

---

33. Tagore, "Sāhitya Shristi," 10, no. 341; translations are mine. Venkatarama Raghavan argues that the concept of Sahitya had a grammatical origin. It became a poetic concept even as early as Rajasekhara [an eminent Sanskrit dramatist, poet, critic]; as far as we can see at present, the *Kavya-mimamsa* [880–920 CE] is the earliest work to mention the name Sahitya and *Sahitya-vidya* as meaning Poetry and Poetics. Even after Rajasekhara, grammatical associations were clinging to the term up to Bhoja's time. Kuntaka [950–1050, Sanskrit poetician and literary theorist], about the time of Bhoja himself, was responsible for divesting Sahitya of grammatical associations and for defining it as a great quality of the relation between *Sabda* [word] and *Artha* [meaning] in Poetry. Sometime afterwards, Ruyyaka or Mankhuka wrote a work called *Sahitya-mimamsa*, which was the first work on Poetics to have the name Sahitya. Afterwards, Sahitya became more common and we have the notable example of the *Sahitya-darpana* of Visvanatha [a famous Sanskrit poet, scholar, rhetorician writing between 1378 and 1434] (Raghavan, "Sahitya," 82).

thoughts and ideas from the Islamic culture and civilization. Minds have met and so have emotions and notions. And to add to the turbulence of the confluence we have thoughts coming from Europe and the minds interacting in these crossroads have woken up to change, surprise, enchantment, and enthusiasm. Literature, Tagore observes, unveils in newer forms when faced with such transcontinental transactions: it is here that the world of *sāhitya* (literature) is formed. Tagore observes that "watching a writer from a compulsive proximity foregrounds a relationship between the writer and the writing as if we begin to think that the glacier Gangotri is responsible for the river Ganga."[34] Writing and the writer's mind—the impossibility of staying alone—are touched and impacted by other cultures, traditions, and thoughts flowing across borders and affiliations and allegiances.

Such diffusion, fragility at the borders of thinking, and athletic plastic imagination throw us into the midst of the "event" of plastic literature. Following Sandra Lee Bartky, we see two formations: one is the "horizontal" being-event that "refers to the meaning of what has heretofore happened, to the way in which Being, which is historical 'in its essence,' has given birth to the epochs of metaphysics," and the other is the "vertical" being-event that "refers to the ways in which within any epoch beings (*das Seiende*) come to be the beings they are." So on that note of explanation, horizontal being-event is about the "varieties of world-disclosure" and the vertical being-event is committed to the "modes of world-disclosure."[35] The world of *sāhitya* is "there"; so before we see a text or a piece of plastic as belonging to a culture, a particular background, a relational context, and a timescape, the world of plastic and plastic literature precedes our reductive experience counterintuitively. This, for me, contributes to how we see *vishva sāhitya* (world literature), presenting and "presencing" its formation not in isolation or apartness but by living holistically; this is the "opening forth." If the pre-reflective and pre-discursive experiences are pressed into play, then the *dasein* of *sāhitya* works around a "poetic" where mattering and presencing oppose the tyranny of the theoretical (the structured) and the given.

What kind of truth are we exposed to? This brings us to the "imperative" to understand the experience and truth of "uncoveredness"—*sāhitya*'s own world-disclosures, its worldings, and the disclosures we effect; for

---

34. Tagore, "Sāhitya Shristi," 350; translations are mine.
35. Bartky, "Heidegger and the Modes of World-Disclosure," 214.

instance, more than what Tagore as a creative writer says or represents (for a material-aesthetic understanding we can consider the twisted plastic bottle with the inscription "Beijing Olympics" washed ashore in Helsinki), we vector toward what unconceals Tagore, the world of Tagore, the "obvious" Tagore, and his being-in-the world (correspondingly, the obviousness of the bottle and what it unconceals). This is not what Tagore does or can be theorized about and, hence, reduced to explanatory parts and his own constructivism, but Tagore as an existing being that has always been on attendance upon alethic potencies—the uncovering of Tagore beyond our worldly understanding of him within his obvious literary, cultural, and political and existential circulations. Don't we find similar resonances in the uncovering or unconcealment of the *Dinosaur* and the *Lamb*? We thus uncover Tagore (technized) as much as Tagore is always attending an unconcealment and "openedness." Would it then be wrong to say that the *vishva* of *sāhitya* determines plastic literature?

Plastic exhausts us with its unprecedented returns and inexhaustible emergences. It worlds in ways that unfixes our thought, structures of expectations, and predictabilities of identity. For Tagore knowing *sāhitya* is often about knowing the limits of questioning *sāhitya*, which does not mean knowing the points of exhaustion; rather, it becomes a reminder of our inability to question further. The aesthetic of Tagore's *vishva sāhitya* can be found in *sāhitya*-disclosures—the truth establishing system, the bringing-forth as an activity that is both translational and transcultural. The world being of his *sāhitya* is to question the conditions of knowledge-generation; it tries to see literature as "existing," as a phenomenon whose truths await to be discovered and are not always imputed and constructed. This is deeply plastic in nature. The truths of such findings lead us to see the "fundamental" of plastic literature, where the fundamental is not merely about what "is there" but about what essentially survives our investigation—Tagore's unexpended quotient of the "literary," the plastic after plasticization (*the Dinosaur* and *the Lamb* are seen as post-plastic). This brings us to question the finitude of plastic literature as a performance and act: if Heidegger has inspired us to question the very role and dynamic of metaphysics in our thinking and understanding of life, I prefer to extend this to our thinking of the *vishva* (world), *sāhitya*, and the *vishva* of *sāhitya*. If every move, gesture, act, and performance comes with historicity, as has been the idealized narrative of expectation and fulfilment in Western cultures, we are missing some part of the world that *sāhitya* ungrounds and something that *vishva sāhitya*, as I see it, has not been able to realize. If *vishva sāhitya*, through a technology of thinking, permits and promotes certain protocols and procedures of doing and performing, it can also make allowance for certain unmapped categories and experiences.

Plastic "World" Literature

It is through the "invasiveness" that drift plastics produce, alongside the long distances they travel to impact other ecosystems, that we observe a dissemination, play, and intrusion based on plastic-habit. Evelyn Rydz's art-series *Floating Artifacts* (figure 17) demonstrates such "invasiveness" that keeps the borders kinetic. Borders become both "interconnected histories" and journeys across time. Disparate plastic objects from multiple origins and dissemination points, like concepts and ideas in a variety of cultures, invade coast-"thought"-lines to uphold "trans-plastic-habit" within the material-aesthetic continuum of trans-literary understanding. This artifact, as Rydz points out, declares "fragments of plastic debris washed ashore under a dissecting microscope to amplify the tiny out of sight residues of our daily consumption and waste into larger life reflective portraits."[36]

In fact, the artifact gives a "sense of scale and the reflective surface quality." This, we agree, is a part of the invasive politics of plastics and its trans-plasticity. The fluid materiality, with its own points of transgression, brings about the disappearance of prominent lines of demarcation that separate identities of thinking and communities of possession. Admittedly, this builds an "across factor" that escapes a method and obviates a method after one has built a connection with a plastic text. Thinking with Heidegger, I call this the "unconditioned across," the across that is more fundamental than we could ever think out; it has its own tribunal of reason. I claim the power of the "ordinary" and the "obvious" in a plastic text and extend the notion of the "across" through terms that are more fundamental, associative, preconditioned, pre-reflective, and, hence, less settled in the "said" than in the "saying." Working through Heidegger's idea of the world, we encounter a space that is "unintended" and present in an "unprominent way," somewhat out-

---

36. Quoted from my personal communication with the artist Evelyn Rydz about her floating artifacts and certain aspects of "plastic literature."

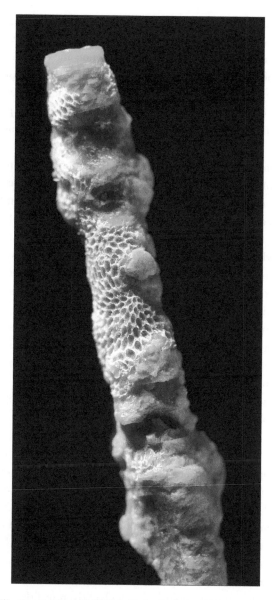

Figure 17. *Floating Artifacts, #8* (2014) by Evelyn Rydz. Used by permission of the artist.

side the conscious formalization of understanding and thought. So what stares back at us is the articulative difference between *vishva sāhitya* and what is "out there" in the *vishva* of *sāhitya*: representation and constructivism need not always find their way in the ways of the *vishva* of *sāhitya* (plastic and plastic literature exceed their states of being and inhabit the "unintended" zones of existence). It is here that the plasticity of literature lies and plastic literature exceeds the possibilities of an institutionalized world literature—more appropriately it may be called plastic "world" literature.

It has been observed that "the size distribution of floating plastic debris point at important size-selective sinks removing millimeter-sized fragments of floating plastic on a large scale. This sink may involve a combination of fast nano-fragmentation of the microplastic into particles of microns or smaller, their transference to the ocean interior by food webs and ballasting processes, and processes yet to be discovered."[37] The sink (shore deposition, nano-fragmentation, biofouling, and ingestion) and dissemination—effected through photodegradation, embrittlement, and fragmentation—are, at once, estimable and unmeasurable, indexable and invisible: this is a clear indication of the "micro-epistemes" (the material-aesthetic of micro-plastics) that literature in a globalized world generates and transfers, possesses and produces. Plastic debris—conceptual debris—reaches the remotest corners of the planet and can also be found in uninhabitable islands. There is no human, yet there is plastic; no agency, yet the literary; no anthropogenic subjectivity, yet object-subjectivity. The nano-degradation of plastic, unquantified and dispersive, makes for plastic ingestion, which, I would like to argue, is both conscious (gastrointestinal obstruction and sorption processes in the form of hydrophobic chemicals) and unaware consumption (coming through, for instance, by fractal process). Such dissemination and transference are unavoidable and voluntary, a poetics of submission and seduction. Micro-plastic fragmentation is progressive and inevitable. So plastic discarded is plastic left unguarded; and its post-representation is possible in forms both visible and invisible. More than biofouling and defouling, it is the sink through ingestion that interests me—its complexity, pervasiveness, diffusion, and contamination. Whether it is in the epipelagic zooplanktivorous fish or small mesopelagic fish, plastic-accumulation through losses and ingestion have their

---

37. Cózar et al., "Plastic Debris in the Open Ocean."

sway and operation, metonymizing the "literary sink," the scopic and synchronic formations.

Plastic immanence is "untimely"; and, incidentally, compared to land fill, haline environments and "cooling effect of the sea" make degradation require "very long exposure times"; "because plastics become fouled by marine organisms relatively quickly, the debris may also become shielded to some extent from UV light, and the persistence of this debris was recently illustrated by accounts that plastic swallowed by an albatross had originated from a plane shot down 60 years previously some 9600 km away."[38] This "delay" and long hours of *exposure*, slow-time moments of long-awaited but inevitable manifestations, as instanced through the albatross that swallows plastic, speak about the complexity of the *materialization*. Significant concepts or congeries of ideas, such as plastic, have a "fragmentation time," which, in a way, explains how they might "sink" and "unmineralize," *trans*-form, and cultivate their "exposure." It is interesting that all plastics (like all ideas) figure out a "chemistry" with their environment or milieu: some are brittle, some photo-degrade with time, some suffer abrasion, some stay unchanged for years together. Similarly, ideas shore up along cultural coastlines; concepts accumulate through centuries and grow their strandlines. Plastic has brought alternatives of use but also levels of anxiety and opposition. Plastic recycling is a fraught and vexatious process as much as ideas across generations find profound opportunities for recycling in modernist-postmodernist aesthetics with varied affect and efficacy—for instance, ideas of Realism, Romanticism, mimesis, sublime, affect, and many others. Plastic recycling offers a serious challenge to the process of "sorting" of plastic waste. Plastic world literature, correspondingly, finds itself exposed to protracted duration of work and effects, to "disposition" and "exposure time" that eventualize in the wake of a sheer processual complexity. Literary-degradation has its own albatrosses in which the show of the "plastic-literary" takes years of process-time, "deep time," the overlapping of sociopolitical factorization, cultural fractality, and a "sink" that is never immediate and abrupt. Here the "plastic litter" is my "literary litter." Plastic literature hides its own innocence and perversity, approximates planetarity and "planetary geoculture":[39] this is no simple cosmopolitanism and

---

38. Barnes et al., "Accumulation and Fragmentation of Plastic Debris in Global Environments."

39. Elias and Moraru, *The Planetary Turn*, xiii.

globalization. More than how we understand world-comparative literature, plastic literature has its webbed and wide immanence in relationality, invisible points of connect and contact, and unprecedented networkism.

Plastic changes its shape and appearance as part of an accelerated network of forces; but it often resists and builds a transformation that is slow and in repose. Accelerated actions trigger rapidly changing conditions demanding quick adaptation and adjustments. Accelerations may be privileged over the "reactive" drive to slow things down permanently, and such upped pace produces greater democratic possibilities, experimental modes, and experiences to understand "deep pluralism" through time as becoming.[40] This acceleration is one form of plasticity that is a product of repose, a slowness, as it takes its own "time" to temper and judge the impulse to act and transform by pausing to evaluate things more; and, often with more time the acceleration produces the moments of "repose"—the time when a thought withdraws and allows itself to be rethought despite considerable pressure on it to be changed and transformed. It has often happened that this repose in thinking has generated a fresh desire in it after it has denied succumbing to the pressure of changing immediately. And inaction (material-aesthetically the non-biodegrability of plastic for a very long duration) conceals unexpected and possible alternatives of emergences and manifestation in the future. The inaction is the agency of slowness, the "slow reading" that plastic world literature generates. Slowness acknowledges the representational precision and pervasiveness of our post-coloniality, the discourses that contribute to its making and also the gradual fragmentation, the interrogation, the challenge of its institutionality, ethics, and prejudice. It is intriguing to observe that slowness breeds complexities more: if speed brings its own unexpected connections, slowness *eroticizes* as plasticity produces sedimentations and sublations, associations and disseminations and formations and decompositions.

Homi Bhabha considers speed—"acceleration and immediacy"—as instrumental in rendering the world "one-dimensional and homogenous." He insists on close reading, on slowness for the "slow pace of critical reflection resists processes of totalization—analytic, aesthetic, or political—because they are prone to making 'transitionless leaps' into realms of transcendental

---

40. Phillips, "Bend, Engage, Wait, and Watch," 98.

value, and such claims must be severely scrutinized."[41] Plastic literature builds its own slow challenge against all forms of totality. This is not to put all kinds of fixity of understanding in jeopardy; but allowing slowness, as Bhabha argues, to become "a deliberative measure of ethical and political reflection," critical inquiry maintains "tension" rather than resolving it.[42] Such tension unblocks points of thinking that might not come through speed and quick jabs of thought; it may need certain changes in our conditions of existence—cultural, religious, and social—impinging on our methods of expression and symbolic forms of articulation. Both the *The Waste Land* and the *Skeleton* or *Dinosaur* and the *Lamb* instantiate such a tension. Plasticity on this note builds ambiguity, which is essential to any form of reading.

Jean-Luc Nancy and Aurelien Barrau have argued that the world "is entering into a movement of indefinite expansion, both on a 'cosmic' scale and in our methods of knowing and acting on it and within it," becoming, in the process "the crucial point where all of the aspects and stakes of sense" in general are "tied together."[43] Sense-making, rather sens-ing, ensures our being in the world, authentically and engagingly, and also secures and drives us into a point above such world-embeddedness to create a different world-meaning. There is what I call a "becoming aesthetic" or a post-aesthetic to world literature comprehensions. This an affective and aesthetic attunement to a world or worlds that reveals in a moment, something that grows with time, eludes our conservative understanding, and discloses to us meanings built through slow evolution, conceptual embrittlement, and repose; such disclosures speak of a worlding, which Kathleen Stewart sees as a being about how "events, relations, and impacts accumulate as the capacities to affect" and be affected as well,[44] Here we encounter another dimension of empowerment in plastic literature that finds strength in being compositional, affective and "combinatric."

As part of the *vishva* of *sāhitya*, the potencies of unconcealment, and the poesis of literary sink, is plastic world literature both a being-in-world and "attunement" (*Stimmung*)? It is interesting to note, following on Heidegger's notion of the "being-complex," that a text written within *desh-kal-patra*

---

41. Bhabha, "Adagio," 377.
42. Bhabha, "Adagio," 375.
43. Nancy and Barrau, *What's These Worlds Coming To?*, 1–2.
44. Stewart, "Afterword," 339.

(context-time-identity) is meaningful only within a relation with others (*sahit*): the being of a text is in the complexity—rather, plasticity—that it builds with others—"a formal or transcendental notion in that it refers to the structure of any possible experience of being-in-a-world."[45] Literature must actually *enter the world* to know how far humans can find their *kinship* in the world, and to what extent they can realize truth. It will not do to know it as an artificial construct; it is a world in itself. Its essence exceeds the individual's grasp. It is in continuous creation, like the material-plastic universe itself, but in the innermost core of that unfinished creation is a perfected ideal that remains unmoving.[46] If *sāhitya* enters (from the Old French *entrer*, meaning "enter," "go in"; "enter upon," "assume"; "initiate") into the world, then it must be coming from a world of its own or worlds affiliated to the writer, his times, his context, tradition, and, finally, a world beyond his own comprehension and construal. Plastic literature's entering then is about worlds coming together, initiated, assumed, and getting into negotiation and play, into forms of expression and aesthetic matterings. So here is the lifeworld that it builds with the world-being: the patterns of disclosures, or the levels of unconcealment that the self and the other in their complex turnings and returnings construct and in which they inhere. Not that such a lifeworld denies the essence of history or historical world-making; rather, world-making as forms of "entering" unfixes worlds of understanding and performs its own disclosive acts of expressions.

We *expend* our lives on plastic; we *expend* our critical-literary thinking on the literary-totality. On both sides of the performance—plastic and plastic thinking—we expend to regenerate. Plastic world literature, in that sense, finds its home in world-disclosures that qualify as a kind expenditure that hardly thinks of losses. Expenditure of this nature reendows the self, brings the self back to thinking about itself, where staying within is reaching out for the world(s) without. Whether through *The Waste Land* or the poetics of cannibalism this expenditure is about "living" in the *sāhitya*-being—its truths (historical and political), happenings (sociocultural), event as happening, and disclosive power (meaning formation). So plastic world literature embeds in "incommensurabilities"—impediments and challenges to think out the *sambandha* ("relation"; it is used both in the sense of context and relation) between the self and the other and in the kinship with the world. It is an

---

45. Bartky, "Heidegger and the Modes of World-Disclosure," 213–214.
46. See Tagore, "Visva Sahitya," 289.

expansion of the circle, the desire to manifest one's self in the other. As Tagore writes, "we see our own character manifest in many people, many nations, many eras, many incidents, many varieties, and many shapes."[47]

Plastic, as anthropogenic debris, can exist and manifest in a variety of forms and states: "Many plastics are buoyant and remain so until they become waterlogged or amass too much epibiota to float,"[48] macro and mega debris, pellets that are ingestible, in sediments. Other seasonal factors include variation in the position of water fronts, the intensity of currents, swell, winds and upwelling, which influence both the distribution and densities. These factors correspond to the variance and valence of the "literary litter" that exists across cultural coastlines, local "currents" of thought and practice, remote and distant territories of availability, spatial scales of emergence, tidal flow, densities of engagement, surprise and strength of dissemination and accumulation: "plastic strands" generated are the material-aesthetic for the conceptual strands in our doing of literature and critical thinking. The movement and *travel* of the plastic litter resonate meaningfully with the state, scale and sink of the "literary litter." Describing the behavior of plastic ocean litter, David K. A. Barnes, Francois Galgani, Richard C. Thompson, and Morton Barlaz explain that

> bottom debris tends to become trapped in areas of low circulation and high sediment accumulation in contrast to floating debris, which accumulates in frontal areas. Debris that reaches the sea-bed may already have been transported considerable distance, only sinking when weighed down by fouling. The consequence is an accumulation of plastic debris in bays rather than the open sea. Some accumulation zones in the Atlantic Sea and the Mediterranean Sea have very high debris densities despite being far from coasts. These *densities* relate to the consequence of large-scale residual ocean *circulation patterns*.[49]

Like plastic litter, which manifests as visible and invisible, literary litter has both a visibility and an invisibility paradigm attached to it. Nurdles (the plastic particles used to manufacture plastic products) contaminate and are fundamental to plastic lit(t)erization.

---

47. Tagore, "Visva Sahitya," 281.
48. http://www.seaturtle.org/mtn/archives/mtn129/mtn129p1.shtml.
49. Barnes et al., "Accumulation and Fragmentation," 364 (my emphasis).

Plastic literature has its own *dharma* in travel, estrangement, and nexus. This is *sāhitya* in its encompassing solidarity—in "comradeship" as Tagore has emphasized. Speaking of Goethe and *literature-monde*, Typhaine Leservot argues that "despite his initial desire to break up the fixed canon of the classics by daring to suggest that it was possible to admire contemporary authors, Goethe's world literature remains an elitist concept which favors the literary production of certain nations over that of others (France over Germany), of certain periods over others (the ancient world over the modern), of certain genres (poetry rather than the novel) and of certain readers (those from the elite classes rather than from the lower classes)."[50] For Tagore, *sāhitya*— plastic world literature—is certainly not hierarchical; rather, he admits to "interference" in all forms of *sāhitya*—interference in forms that are cultural, rhetoric, political, linguistic, epistemic, and national. Literary expression builds its connection and also stays connected in ways that are too subtle to methodologize and express programmatically—the microplasticization. Anachronic and anarchaeologic,[51] the *vishva* of *sāhitya* is not a cluster of texts predetermined through certain protocols and patterns but a meaning-making process, a plasticity, a kind of nonreduction that is poetic and aesthetic at the same time. Uniquely, if Tagore brings us before world-comparative poetics, for me, it leads us into plastic literature.

It is interesting to observe the "estrangement" that plastic lit(t)erization builds: writing and expression are one's own, and yet the greater currents take one's work outside of what one intended to establish and formalize. Far more profound and intricate than what conventional premises of comparative literature would allow, the very idea of plastic literature is a mode of relationality—the sight of the *Dinosaur* and the *Lamb* and the reading of *The Waste Land* produce estrangement and attachment simultaneously. Such a movement is the plastic text's lifeworld that is always an unintended victim of disclosive release. As Pheng Cheah notes,

> Structurally detached from its putative origin and that permits and even solicits an infinite number of interpretations, literature is an exemplary modality of the undecidability that opens a world. It is not merely a product of the human imagination or something that is derived from, represents, or

---

50. Leservot, "From Weltliteratur to World Literature to Literature-Monde," 41.
51. See Zivin, *Anarchaeologies*.

duplicates material reality. Literature is the force of a passage, an experience, through which we are given and receive any determinable reality. The issue of receptibility is fundamental here. It does not refer to the reception of a piece of literature but to the structure of opening through which one receives a world and through which another world can appear.[52]

This is the being-event of plastic literature whose totality is formed through *sambandha* that is achieved through structures of opening. As a "network of nodal notion," according to Vilashini Koppan, the world is "both a way to name a totality (the network) and a point of location, a placing or emplotment within the totality"; it is about occupying "two places simultaneously, seeing both from afar and close up, zooming in and zooming out, looping in and looping out."[53] Plastic literature inspires such becomings. The world in plastic literature is powerfully subjective and extra-subjective too: the extra-subjectivism comes from one's helpless submission to invisible forces of textual transmission, the inability of the subject to control the forces that determine the future of a work—the text's own world-disclosures. So the world of plastic literature builds its resources and expanse through a new order of production both through its writers and the writing, as every writing becomes its own rewriting—a kind of co-occurrence and co-performance—through its transference and transmission in the global circulation of literary and market capital. Here I would like to see estrangement as attachment: attachment as mesh, as nexus. Plastic world literature needs to look into promoting good and effective cosmopolitanism and also "estrangement as interconnectedness"—a together-apart syndrome—through informed and productive ways of understanding and judgment.

The estrangement is, in fact, translational: a work written within the confines of a nation, a community, builds its own *rasa* outside the intentions and commitments of the writer. If a work relates to the circulation of writing outside oneself, the *rasa* of the work is relational. All expressions are translational; "everything is translated," notes Bruno Latour; "we may be understood, that is, surrounded, diverted, betrayed, displaced, transmitted, but we are never understood *well*. If a message is transported, then it is transformed."[54] The travel

---

52. Cheah, "What Is a World?," 35.
53. Coppan, "Codes for World Literature," 107.
54. Latour, *The Pasteurization of France*, 181.

in plastic literature through translation is also the travel through transformation; for me, besides worldiness, there is a world outside the text that sponsors the worlding, the ambiguity of its reception and reading, the politics of its global and transcultural flow and the power politics of language and dissemination. Caught in symmetrical-asymmetrical translation, geo-critical spatialization, itinerancy, and transcultural semiosis, plastic literature, thus, spells out a *totality*; it speaks of a formation that does not allow interpretive conquering but has its own ways of manifesting and reordering. Totality forms additively with the worlds of a variety of literatures emerging from a variety of cultures, times, and places. But this aggregatory formation challenges itself every time one tries to conceptualize its existence. By being plastic, the literary-totality need not be misconstrued as totalitarian, for the enfolding and unfolding continue with a variety of manifestations and reorderings.

John Dahlsen's *Gyre* series (figure 18) brings us to the plasticity of space as we realize how plastic waste travels into the existing gyres to enlarge and intensify them. The *Gyre* artwork shows the miscellany and divergence of material incorporation; and it makes visible an ever-increasingly complicated pattern and formation as materials flow in from disparate directions to generate an entangled circulation of material-mass defined by density, concentration, mobility, and scalarity. The *Gyre* looks a totality, but it is not an immobile congealment around a particular point of emergence. The buoyant plastics in unimaginable heterogeneity (its center, estimated to be one million square kilometers in size, may contain a trillion pieces of plastic[55]) "travel" from across Asia and North America riding the sea-currents to silently dilate the seemingly fixed body of material mass.

"There are five major subtropical oceanic gyres," observes Elizabeth DeLoughrey, "which, on their surface alone, have accumulated upwards of 40,000 tons of microplastics. But their volume and depth are much greater, and harder to fathom."[56] Their foundations are in irresistible mobilities, wind and sea currents, an undergirding swirling force, and re-formations. They destabilize themselves in a continuity and convulsion that correspond with the "deconstruction" of the canonistic principles of world literature, making for scalar and volumetric changes in understanding the nation, its

---

55. "Evidence That the Great Pacific Garbage Patch Is Rapidly Accumulating Plastic"; "Plastic Ocean." See also Oliver Milman, "'Great Pacific Garbage Patch' Far Bigger than Imagined."
56. DeLoughrey, "Gyre."

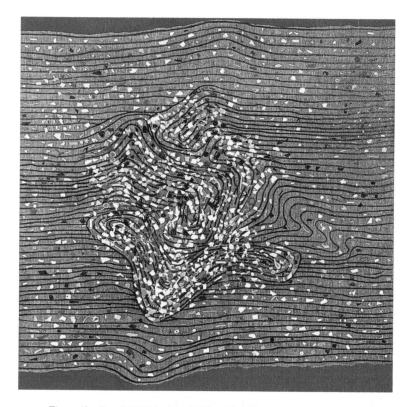

Figure 18. *Gyre 1* (2020) by John Dahlsen. Used by permission of the artist.

borders, and imperialistic politics of literary framing and manipulative ideological drift and discourse. The unfathomable and ever-widening gyre analogizes the worlding of plastic literature with its uncanny verticality and inter-determinancy of scalar sweep. It is where "desire is mobilized and set into circulation, and where our 'projections' about others are negotiated."[57]

Totality is in the *across* and lubricates the idea of space and place in the *vishva* of plastic literature; it disfavors canonicity and elitism, which are strategically and preferentially inclusive of works across national/regional literatures, and refuses to be daunted by the immensity that informs the "quantitative approach" to reading world literature. Plastic literature sets itself up as a "happening" across cultures and times, questioning the fluid routes and

---

57. Bahun, "Politics of World Literature," 373.

rootings of a work. We expend and something grows at our expense—a fluidity on the other side of consciousness, a developmental rhythm effected through our relentless investments and unabated expenditure on our desire and habit-abilities. The gyre grows as much as plastic world literature with its own forms of *expenditures* through transculturality, post-national constellations and the unhinging of the local-global bind. The mobile and ever-enlarging "totality" of the gyre, its changing densities, flow-patterns, variations in material influx, degradation, and fragmentation effected within, make it a material-aesthetic phenomenon for rethinking the "plastic literary."

Plastic literature has, to borrow from Michel Serres, a "flow"—the kinesis of vortices as they interact and raise possibilities of fresh emergence. Serres explains that "the vortex is unstable and stable, fluctuating and in equilibrium, is order and disorder at once, it destroys ships at sea, it is the formation of things."[58] More powerful and potent than world literature, plastic literature is never without its own gyres—"ensemble of fluencies"[59]—and creates its own geometricizations and angularities. The vortices complicate representation, or rather, signification. In fact, they complicate "sense": as Ignaas Devisch writes in his critique of Jean-Luc Nancy, "The world is structured as sense and sense is structured as world."[60] Plasticity is left as "coming to presence," the power of arrival in the arrivant and the arrived. The language of a plastic text, whether *The Waste Land* or *Waiting for Godot* or *Finnegan's Wake*, is fragmentation and possibility, making for "transimminence." Plasticity as flow enunciates a discourse in place (the plastic patch or text patch as a site of formation, insemination, and dissemination), concealment, and losing of places. Threatening to self-exceed itself always, plastic literature speaks of a performative "coming into being": concepts, images, understanding, ideas as they come into being across traditions and cultures in a figuration of sense: it is the "opening." The gyre declares a negotiation of a world as moving *across* worlds—social, political, religious, and cultural. Negotiations of such kinds are close to Nancy's "mondialization." François Raffoul and David Pettigrew clarify that mondialization "maintains a crucial reference to the world's horizon, as a space of human relations, as a space of meaning held in common, a space of signification or of possible significance," while "globalization is a process that indicates

---

58. Serres, *The Birth of Physics*, 30.
59. Serres, *The Birth of Physics*, 58.
60. Devisch, "The Sense of Being(-)with Jean-Luc Nancy."

an 'enclosure in the undifferentiated sphere of a unitotality.'"[61] Defying unitotality or transcendent singularity, plastic literature claims a shared world with
relentless mediation and transmission of meaning, ideas, and understanding.
On the axis of mondialization it generates and projects world(s) that are analytical, constructive, often unknowable, processual, and encompassing.[62] Can
Waiting for Godot, Ulysses, The Waste Land, Midnight's Children, and others be
instances of world-widening gyres? Bioplastics stay suspended under various
stages of degradation; concepts and ideational linkages are suspended particles
in the plastic "literary" gyre, which, often, gets spit out into certain accumula-
tive forms with varieties of effects and modes of consumption (say, reception).
The gyre self-exceeds and keeps on refiguring itself as much as the plastic literary that initiates different formations beyond its established core of existence.
The gyre plasticizes; islands of plastic-thoughts form and extend; the conceptual soup announces identity, but not without an inherent connective and ac-
cumulative strength and energy. The "folded waters" are folded identities: we
own plastic in a continuum, indefensibly transmissive and transformative, as
much as we live in a literary-totality. This totality is plastic-restive and plastic-
transitive in a kind of inexhaustible plenitude.

If knowledge is produced in congruence (plastic literature never claims
absolute outsiderhood to such possibilities), knowledge owes to asymmetry
of relations as well. Plasticity in thinking comes from the spacing of the world
in what Nancy sees as the "dispersions of origins, their dissipation and their
prodigality."[63] Plasticity helps us to build literary-comparative judgment as
all cultural formations in their micro- and macro-dimensions exhibit the "excess of the finite." Tagore, Eliot, Eco, Borges, and Alberto Manuel are finite
in their own ways, but not without the distant co-sharing of spaces—
mondialized in plasticity. However, being plastic is not merely about "transit" but about maintaining identity as well. Plastics face biodegradation, are
incinerated and recycled as a kind of a change of identity that does not render them out of existence. And, much in the same way, literature in the globalized world survives prescriptions and perversions of transcultural thinking. It is significant to know that plastic mobilizes its own biodegradability
and resists its own biodegradation (among modes of degradation plastic can

---

61. Raffoul and Pettigrew, "Translators' Introduction," 2.
62. Raffoul and Pettigrew, "Translators' Introduction," 28.
63. Nancy, *Multiple Arts,* 177.

be exposed at the surface to dampness, bacteria, enzymes, and other forces; favorable landfill or composting systems can effectively contribute to biodegradation, and, also, plastic is generated through biodegradable additives). But the wholesale breakdown of most plastic is not possible; it mutates, morphs, and metamorphoses but resists extinction. On that material-aesthetic note, plastic literature spells out its poetics of dissolving identities across cultures and national borders with the reminder of preserving certain identities that no form of "worldization" or globalization can undermine. Post-plastic is a state in which biodegradability results in forms that do not resemble the plastic we know and use but are an extension of plastic—a rematerialization that has its own politics and thingifying abilities. Plastic literature pitches its faith in such rematerialization of meaning and manifestation.

Rematerialization and reconfiguration bring us before a plastic planetarity with a different "smoothness" and movement, a different gradient and affordances. If plastic has built its own "relational potenza" hypertrophizing the idea of the global, plastic literature is our fresh poetics for conflating the "imperative" and "indulgence" in literary-critical thinking. The plastic-literary-litter plays up the game of "unworlding" literature, diversifying the transnational spaces and cross-border appropriations in a restless parataxis, re-premising the local-global divide and the center-margin problematic. So the ability and disability with plastic is coextensive with our reading of texts. Plastic spurs the imagination; ingenious texts stir as well. Plastic pains; texts can be desperate too. Plastic is (un)ease; texts genuinely are. Plastic is ambiguous in pleasure and pain; texts are never short of such ambiguity. Plastic changes its meaning as it progresses; texts do the same. Plastic charms and baffles; texts act similarly. Within such an ambit of plasticism, we engage with two mariners, two kinds of sea, and two kinds of albatross: the migrancy, immanence, and *seepage* in reading and formations of the plastic imaginary by looking into Coleridge's "The Rhyme of the Ancient Mariner" and Nick Hayes's *The Rime of the Modern Mariner*.[64] The plastic-worlding of the two poems in their visibility (equivalent to the plastic that we get to see) and imperceptibility of meaning and entanglement of expression (the plastic that exists and builds unseen) are different: Coleridge

---

64. Hayes, *The Rime of the Modern Mariner.*

wrote a "voyage" poem benefitting from Wordsworth's reading of George Shelvocke's *A Voyage round the World by Way of the Great South Sea* (1776), which had the episode of the albatross-shooting, an adroit and typically Coleridgean mix of geography and supernaturalism, the historical and conceptual correspondence with James Cook's voyage, Francis Drake sailing through the Magellan Strait, Frederick Marten's *The Voyage into Spitzbergen and Greenland*, Alexander Dalrymple's *A Collection of Voyages Chiefly in the Southern Atlantick Ocean*, and mining expedition narratives (Martin Frobisher's in the 1570s, John Davis's rediscovering Greenland in the 1580s, Arctic voyages through the seventeenth and eighteenth centuries). Hayes, on the other hand, demonstrates a green-and-blue consciousness that is deeply distracted by plastic pollution, oil spill, relentless and inexorable sea-contamination; it is a hugely conscious poetic mind that engages with ecocide, waste, filth, and other dimensions that close out the anthropocenic circle. This is the "sink" of the literary that settles into the depths and recesses of a text; it is a contribution that is imperceptible in its apparentness but deeply meshed—a kind of meso-plastic litter. The sea, for Coleridge, was imaged out of his readings, his engagements with travel writings, and his esemplastic imagination that did not fail to have the icy and "copper" sea and the Albatross overhead. Hayes's sea, on the other scale, has a reality that is portentous and perilous—the reality of a plastic sea and the Albatross with intestine-gutting plastic in its stomach.

In Hayes's poem, the "modern" mariner's ship left the "harbour gracefully" (line 39) and 'slid through soapy lotion" (40). He watched the "flotsam on the wash" (44) and "with little else to do (41)"

> So took my gun to practise aim
> And shot a bottle or two.
> But shooting plastic sitting ducks
> Soon failed to entertain;
> I lowered my gun . . . and saw a bird
> And picked it up again. (42–47)

But this time he aimed at a "creature" that "soared high above" (48) and brought it crashing upon the deck. "It was *just* an Albatross (53)." In the wake of the shooting, after a peaceful afternoon, trouble broke loose with the "chugging engine" (58) dying, the wind falling still, as the ship "slid into" the "North Pacific gyre" (61):

> A desert of subtropic brine,
> A lifeless Maelstrom
> A stifled silence cloaked the sky,
> Bereft of seabird song. (62–65)

Berated by his crew, the mariner looked across the sea and, to his utter dismay, found themselves beset by a "wash of Polythene" (77), "swathes of Polystyrene" (78), "bobbed with tones of Neoprene" (79), and Polymethyl and Methacrylate (80). The slimy creatures of the ancient mariner have been replaced by Tupperware and bottletops, bottled bleach and tires, and a variety of plastic detritus creating a "funeral pyre" (84). The Albatross transcorporalizes from a bird to plastic in a kind of transvaluation where the good omen of the bird for Coleridge changes into the sinister truth of plastic death and dismay for Hayes. The albatross is the modernist dilemma of "finitude": a bird or a phenomenon that does not wait for independent agency or will to execute an engagement. It is an event in "slow violence," for what we do to ourselves happens to it as well. Our dying reflects on its dying, too—an Otherization achieved through "dying." This is the modernist predicament of "death-in-life."

Is our ecocidal fate an albatross syndrome? Whereas Coleridge's sea sees ice-crack and growl, roar and howl, to trap a ship, Hayes's sea finds itself in plastic soup, messy and colorful, deceitful and seductive, killing and forbidding. Here the mariner does not need to shoot the bird, for the bird stands shot by a different kind of human agency: humans' discovery of a substance about which they do not know what to do to avoid being dehumanized. Both the mariner and the albatross share a plastic identity—a new community in plastic-sharing, a plastic bondage. One of the interesting things to observe is how the albatross, the mariner, and the sea are intermeshed in plastic—a transubstantiation that sees a valuational rephrasing of the Coleridgean expression of "blue sky bending over all" changing into "all bending before plastic." It is plastic, plastic, everywhere, not a drop to drink sans plastic. The modern mariner, as it were, has shot himself down with a plastic crossbow.

If Coleridge's ghost stood on the deck, Hayes's ghost declared its existence "nine fathom deep" (88), manifesting in nylon nets, acrylic, foam, and polymers as they netted and degraded almost all sea creatures—birds and fish. The despairing sailors, bone dry in their throat and flummoxed in the doldrum air, hung the carcass of the albatross around the neck of the mariner: "I bowed to take my Cross" (106). Slung around his neck, the bird was rotting with a "nylon gauze" tangled in its chest: "its kidneys, withered by the sun, were strangled by this mesh / A Plastic bag hung round its bones and decomposed like flesh" (144–146). This is a new Albatross: a different *nature*, anti-Romantic.

Albert Ross, the massive albatross in Nicola Leigh's children's book *Blue Spaghetti*, can bring only plastic back from the sea for his chicks: "No red fish could be spotted. . . . New food had arrived that was all blue and stringy. He scooped it up to try, and named it 'Blue Spaghetti.'"[65] In fact, in as remote a location as Midway Atoll, researchers have found that the stomachs of Laysan albatrosses contain a mix of organic matter (charcoal, candlenuts, walnuts, squid beaks, wood, sponges) and inorganic matter (plastic caps, bottle fragments, tubes, broken pieces of toys, polyethylene bags). This diversity of "stomached" ingredients raises questions as to the ingestion, transference, and contamination processes: "It is a behavioral characteristic of albatrosses (as well as some other pelagic birds) that items (food or other things) are rarely ingested on land. It is therefore not surprising that nearly all of the foreign material found in the dead birds floated. In consideration of the pelagic feeding behavior of albatrosses and the above observations, it is concluded that the plastic and pumice items were picked up at sea by the parents and then passed on with regurgitated food to the young."[66]

Discarded plastic, in forms unconceived and unforeseen, can reach these remote and distant lands to infiltrate the systems of unsuspecting creatures. Dead albatrosses are evidence of how our creation has surpassed the creator to build a freedom and independence to challenge the anthropos. Here the anthropos stands exceeded as the discovery, dissemination, and damage of plastic recursively compel the individual to rethink his or her limits and the sense and possibility of limits. If plastic was the surprise of the anthropos at the beginning of the last century, plastic today is the surprise that has met him or her from the other side. The counterinsurgency is the uncanny dehumanizing process of writing back the history of planet life, the planet-discourse. The world-view now is the plastic-net. This plastic-net also brings up the argument

---

65. Leigh, *Albert Ross, The Albatross.*
66. Kenyon and Kridler, "Laysan Albatrosses Swallow Indigestible Matter."

centering on the complex currents between two supposed points of journey: plastic production and the intestines of the albatross. Coming from a variety of complicated sources, routeways, and flight-ways[67] plastic debris floating in the sea seduces these birds to "pick" plastic-as-food and food-as-plastic. This is the complexity and entanglement of ingestion, as a game, a necessity, and a strategic selection. The "ecological wounding" here is the point of aesthetic-literary formalization: the plastic lit(t)erization.

Which albatross, the one that has plastic or the one without it, is more real? The hyper-objectified sea provokes a subliminal thinking, a "being-quake" (in the words of Timothy Morton). The plastic in the stomach of the albatross petrifies the mariner into thinking the invisible hyper-objects that defy the "mathematization of knowing." Morton points out that the "overall aesthetic "feel" of the time of hyper-objects is a sense of asymmetry between the infinite powers of cognition and the infinite being of things. There occurs a crazy arms race between what we know and what is, in which the technology of what we know is turned against itself. The arms race sets new parameters for aesthetic experience and action, which I take in the widest possible sense to mean the ways in which relations between beings play out."[68] It is the relationality that, in an anti-Romantic way, has brought about a de-individualization where nature is seen to construct itself. The Romantic imaginary is not exclusively Coleridge's secondary imagination alone; it is also the "nonlocality, temporal undulation and phasing" that hyper-objects have brought. The sea, its waters, creatures, birds, the sky, the breeze are interobjectified, en*mesh*ed. "Every interobjective space," argues Morton, "implies at least one more object in the vicinity: let us call this the 1 + n. Writing depends on 1 + n entities: paper, ink, letters, conventions. The human anthropomorphizes the cup and the cup cup-omorphizes the human, and so on. In this process there are always 1 + n objects that are excluded."[69] Plastic can *stick* to us in ways and means that are unfathomable, in scales that are often immeasurable, in moments that are, most often, unguarded (Rydz's artifact articulates these aspects meaningfully).

This hyper-objectification, rather, interobjectivity, stares back at the modern mariner in his wounded sensibilities. The sea appears uncanny as ob-

---

67. Dooren, *Flight Ways*.
68. Morton, *Hyperobjects*, 22.
69. Morton, *Hyperobjects*, 89.

jects external to the common culture of sea-habitation (a "confetti of plastic bottle lids") emerge to project a disturbing "dwelling." Plastic-denaturization deludes sea creatures into trick ingestion through its color and shape and appearance; plastic cheats its way into humans' systems with greater stealth and steadiness. The apparition meeting the modern mariner does not merely metonymize sustainability and survival; it problematizes the whole event of "withdrawal." As an existential affect, eco-aesthetic and material-cultural in nature, the albatross—the "coming together" of Coleridge's avian guest, Hayes's plastic-hurt bird, and the plastic hit Laysan albatross—signals our "staying with trouble" entwined as it is "in myriad unfinished configurations of places, times, matters, meanings." The modern mariner finds himself in Chthulucene, which, Donna Haraway explains, "is a compound of two Greek roots (khthôn and kainos) that together name a kind of timeplace for learning to stay with the trouble of living and dying in response-ability on a damaged earth."[70] If plastic exists in the intestines of the albatross, plastic is inside the mariner's body too. It has democratized our plasti-citizenship. They are Chthonic: "They writhe and luxuriate in manifold forms and manifold names in all the airs, waters, and places of earth. They make and unmake; they are made and unmade. They are who are."[71] Plastic is the mariner's "oddkin." We are in the times of plastic-compost: "We become-with each other or not at all. That kind of material semiotics is always situated, someplace and not noplace, entangled and worldly. Alone, in our separate kinds of expertise and experience, we know both too much and too little, and so we succumb to despair or to hope, and neither is a sensible attitude. Neither despair nor hope is tuned to the senses, to mindful matter, to material semiotics, to mortal earthlings in thick copresence."[72] The sea for the mariner is holobiontic: complex patternings involving multiple sea creatures, manifestations of plastic debris, oil, flotsam, sludge, albatross, mariner and his crews.

We are faced with the thesis of extinction and sustainability within a complicated dialectics of struggle. How can our engagement with plastic and, increasingly, plastic's relation with us, keep up the sustainability game?[73]

---

70. Haraway, *Staying with the Trouble*, 2.
71. Haraway, *Staying with the Trouble*, 2.
72. Haraway, *Staying with the Trouble*, 5.
73. See Griffiths, "Jorie Graham's Sea Change."

Interestingly, sustaining with plastic is what Hayes's modern mariner emotes and exemplifies. Plastic has been bringing a "new order" through an exceedingly anxious and helpless submission, a dreadful relentlessness. Our eating, drinking, cultural and aesthetic constructions, our relationship with the non-human, circumambient eco-poetical materialities are as much *in* plastic and *with* plastic as our presentness, past, and future. Plastic has brought us into interobjectification and in a subject-object divide: the mariner has plastic as much as the albatross, and yet, aghast and pertrified, he watches the hyper-objectified sea growing in its unfamiliarity and disconnect from him. Plastic fraternity and fractality have become simultaneous. If the ancient mariner overcomes the "sliminess" to accept his participation in the life of whatever is alive, the modern mariner cannot help being webbed in the life of things. Interestingly, we are confronted with two kinds of sea: not merely an old sea and a new one, as a sea that is new with submarines and cruiseliners as contrasted with the old, which had more whales and other endangered species. The sea is getting "newer" each day, transforming itself like the forest cover in the land within a "yet-to-be-born" nature that bespeaks our fragility and extinction; it is the gothic art of extraordinary objects coming together through oil spill, plastic litter, and contamination, sound pollution, and a rare marine plunder of egregious dimensions. If the ancient mariner's sea called for his participation in God's world of creation—the state of "blessing," in agape, all within the arch of the blue sky—the sea for the modern mariner revolts and revulses, making participation a helpless surrender, a shudder of being, where the smoky and choky sky rules over all, "both man, and bird and beast." Morton argues that the "problem of human beingness, declared Sartre and Lacan, is the problem of what to do with one's slime (one's shit): 'The slimy is myself.' So ultimately, is sliminess not the sacred, the taboo substance of life itself? One word for this is Kristeva's abject, the qualities of the world we slough off in order to maintain subjects and objects. Ecological politics is bound up with what to do with pollution, miasma, slime: things that glisten, schlup, and decay."[74] The being of the mariner and being of the sea "quake" up into a crooked straightness of a relationship.

Putting Edmund Burke's notion of "symmetry as sublime" to work, I would like to call this syndrome "immanent symmetry," where the sublime

---

74. Morton, *Ecology without Nature*, 159.

is in finding plastic in man, in all biotic nonhuman creatures, in animals cut off from the direct heat of human civilization at remote places on the planet, in near-inaccessible corners of this earth, and in depths of sea unexplored and unfathomed. This is the time of "plastic symmetry" (*symmetria plastica*), and the intensity and imminence of such plastic-symmetricization are sublime. The Burkean paradigms of self-preservation and terror in the construction of the sublime change into the paradigm of "fragility of helplessness" in the face of the emergence of plastic sublime. A withdrawal from plastic cannot change our plastic-meshed existence; if withdrawal comes in the form of abstinence, we are still temporally boggled to figure the phasing out of plastic. This is the sublimity of fragility, a somewhat loosely understood interobjectification whose end is always much more extended and complicit than what we envisage. Plastic has made us reflect on our limits. They are the ceaseless "quake" of plastic—our plastic-beings and plastic career. Plastic made us think; but, now, beyond correlationalism, plastic is thinking in modes of relationality and assemblaging that are beyond human will and cognition, emerging from the realm of the non-anthropogenic.

Plasticity in plastic literature is not a priori only. It undergirds all discourses around *vishva sāhitya*, our thinking around any form of comparative literature. Approximating Mikel Dufrenne's notion of a "world atmosphere,"[75] it becomes an affective quality that is incidental to any aesthetic object, whether a play or novel or poem or art-work, something that provides the coherence and the binding principle. Permeative and intrinsic, it radiates an atmosphere of its own. Just as all texts across cultures and traditions have a world atmosphere to them, so also texts have their own plasticity. Intriguingly drawn to Dufrenne's a priori where the "knowing subject" is not the sovereign and where all knowledge cannot be subjected to its validation, plasticity is often the "objective a priori" something whose meaning is "grasped in experience and instantly recognized."[76] Plasticity as a priori has its emergent conditions in history and culture, and by that argument plasticity as a philosophy comes in two forms where one is the a priori and other is the non-subjective emergence that works outside it. To extend the argument further one may see the "non-subjective" as the a priori where plastic world-comparative literary formations have non-experienced and non-objectified

---

75. Dufrenne, *The Notion of the A Priori*, 178.
76. Dufrenne, *The Notion of the A Priori*, 59.

dimensions through chance and sudden intersecting points—plasticity as the a priori experienced and embedded in culture, history, and society and also the uncategorized and uncalculated. Plastic-world literary formations are not extensions of the "constituted" only; the innate fragility of their constitution enables the constituting of spaces in joy, experimentation, and risk.

Plastic-world literary formations are, hence, a deeply problematic relationality, a kind of symbiotic assemblage wherein the "knottings" and "clottings" of existence are bounded, formal, anthropogenic, intra-active, imminent, and immanent. It is intermaterial sympoesis. Plastic has changed the politics of our skin-borders. We are suddenly and ardently embattled within a new poetics of relationality, immunity, language, means of absorption and ingestion, and what Haraway calls "disease as relationship."[77] This condition is the diffraction that plastic has brought and correspondingly imported the force and potency of the "literary" into our thinking of comparative world literature. By understanding distance that is never without nearness and desire and that is never without disagreement, we are in the midst of the "performance" of literature. This performance ensures a co-optative, coadunative, and co-constructive relationship among disciplines of thought and epistemes. The reading of the two poems as instances of plastic literature resonates with Deleuze and Guattari when they see a book as having "lines of articulation or segmentarity, strata and territories" and also "lines of flight, movements of deterritorialization and destratification." In fact, plastic literature espouses "comparative rates of flow" that "produce phenomena of relative slowness and viscosity, or, on the contrary, of acceleration and rupture." Some lines have "measurable speeds," and some flights are "unattributable."[78]

A trans-plastic rhythmanalysis in ideational entanglements—Hayes is with Coleridge in acts of manifest "rewriting," Hayes and Coleridge negotiate with Deleuze, Latour, and neomaterialist thinking, the benign albatross connects with the Layson albatross, seafaring finds paths of communication with sea studies, aesthetic-religious studies grids with ecocidal blue humanistic thinking—contributes to a "plastic assemblage" of literature. Plastic planetarity projects the "literary"—metaphorically, hyper-tropically, and performatively under the sea, within the sea, and outside the sea. Plastic defies and connects us; it has a predictable pattern to its relationship with humans and

77. Haraway and Goodeve, *How Like a Leaf*, 76.
78. Deleuze and Guattari, *A Thousand Plateaus*, 3.

their world—the planet life—and has its own reason to dissipate and chance about, leaking, leaching, seeping to extend its radius of contact. Concurrently, plastic literature is both about f(n)orming a system of thought and about allowing the form-ability of thinking where the conscious and conditioned coexist with "what happens" and what emerges and defies us non-algorithmically.

The material-aesthetic of the plastic-comparative of the two poems demonstrates a "plastic reading." Plastic exists in transformation—a universality and ubiquity that build in a self-transcendence of its status both as a material and aesthetic. It is a plastic spread that is at once a constant (structure) and a variable (after-structure). The plastic-gyre of understanding the poems from two different points in time and contexts unfixes the contingency, concept, and history of reading. Just as plastic unbuilds itself, the reading submits to a disruption in time, contexts, and transgressive appropriation of concepts. Within the plastic-gyre imaginary, I see a disruption of the chronic, the periodization, for the history and time of one plastic-thought, on certain occasions, is difficult to differentiate from the other. Plastic ephemera drifting into the Gyre from China, Bangladesh, Britain, and Mexico are often difficult to distinguish. But these are not without a structure that comes after the unbuilding; it is a structure that keeps coming to write itself after a continuity in decomposition. The plastic accumulation and drift are structures that trespass and displace themselves to become the *other* without losing their structurality.

Close to Catherine Malabou's "plastic reading," which has the structure of a "structural plastic analysis,"[79] the reading of the poems shows how plasticity is not simply about reckless self-transformation and deconstruction. Malabou notes that "structure is not a starting point" but "rather an outcome";[80] transplastic reading produces a structure after its order of understanding has come to be deconstructed. The reading of the poems is an "outcome," which, like buoyant plastic with its mobility, mutability, and totality, aestheticizes a movement in form-ability. A regenerative reading does not lead to a structure or leave a structure behind as a "remainder" to protect and sustain. Plastic leaves no permanent remainder, for we observe that its degenerative consequence as a material is also a structure that comes after a structure. Plastic reading, as much as the plastic literary, is an *event* (from the French *événement*, happening, coming to form) that launches itself after "reading," as a desired or consequent

---

79. Malabou, *Plasticity at the Dusk of Writing*, 51; see also Malabou, "La generation d'apres."
80. Malabou, *Plasticity at the Dusk of Writing*, 51.

form, has come to an end. Plastic reading and plastic weathering are all about a sense of a dynamic structuralism after structure.

Plastic disturbs; it has turbulence; it is conflictual and negotiatory. The plastic literary embosoms a turbulence, for it is systematic and random, varying its scales of distribution and dissemination and building its own patterns. If plastic has changed our "now," the *literary* working across cultures and traditions believes in the "layering" of the now: the now is not an objective point of separation from yesterday or tomorrow, but an event, a moving now, that has multiple seriality to it—a kind of observer-independent and observer-dependent "hyperplanes of simultaneity."

Plastic li(t)terization gives us a fresh perspective on the "taking-place" within the complexity of time-curve, surface-depth (close to pentimento), visibility of the invisible, and an understanding, Michel Serres–like, of the "static" as entropic. In fact, material-aesthetically, both microplastic and plastic literature are gridded on a "passion at a distance" (in the words of the physicist Abner Shimony).[81] Our thinking and idea-formations are steeped in passion that only distance can generate, resulting in a dynamical transience. The locality and translocality of literature—the plastic in the tap water and in the stomach of the Layson albatross—index the dynamical transience, embed the entanglement (*Verschränkung*) of potentiality and actuality, are invested in propensity and purposiveness and about concrescence and contamination. The fundamentals of any cultural and literary formation, paradigmatic or conceptual, are maximally vibrated through maximal entanglements. This makes plastic literature entropic within a trans-plasticity that is fundamentally five-dimensional—events that exist as particles in unreal time, as waves across history through real time, the spatiality and temporal relativity of their existence, the hidden variables that inflect and influence them, and the transformative power, the implicate and conative potential, that make certain things happen outside our spatio-temporal understanding. The plastic web of the "literary" is complicit in these actualizations. Being tensional in a post-Romantic way, plastic becomes a statement, a negation, aberration, abandonment, uncanny, shock, surprise, rescue, and return. And doing plastic literature is no different from that as it ensures its existence within such a transvaluational performative as a project in motion, incompletion, and failing.

---

81. Howard, "Passion at a Distance," 4; see also Shimony, *Search for a Naturalistic World View.*

Chapter 5

Plastic Affect

Plastic is the silly putty, with which we simulate, then supplant, every facet of
reality, converting all the varied elements of the planet into one common
emulsion. While we sleep, our automatons toil throughout the night,
transmuting everything into a petroleum byproduct that resists bacterial
predation. Our species might openly mourn this phase of our demise, but in
secret we really exalt the power of its genius, marveling to think that, in some
landfill of the future, long after our own extinction, a single crash helmet
might still endure, sloughed off, like the carapace of some alien crab. Our
gewgaws of epoxy resin and nylon fibre do not attest, however, to any advance
in our rational prowess so much as they allude to the breadth of
our cultural tyranny.

—Christian Bök

In June 1833, Charles Darwin requested the captain of the HMS *Beagle* to
delay his departure from Tierra del Fuego because a "strange group of granite
boulders" has stirred his investigative and imaginative energy: "One of these,
shaped somewhat like a barn, was forty-seven feet in circumference and projected five feet above the sand beach," he later wrote.[1] In his accounts of the
voyage of the HMS *Beagle*, Darwin described crystalline boulders of notable
size and abundance near Bahía San Sebastián, south of the Strait of Magellan,
Tierra del Fuego. Influenced by Charles Lyell's reflections on slow, vertical
movements of crust, submergence, and ice rafting to explain drift, Darwin
proposed that the boulders of Bahía San Sebastián were ice-rafted.[2] Darwin

---

Epigraph from Bök, "Virtually Nontoxic," 25–26.

1. Lovett, "Darwin's Geological Mystery Solved."

2. Evenson et al., "Enigmatic Boulder Trains, Supraglacial Rock Avalanches, and the Origin
of 'Darwin's Boulders.'" Darwin's thesis, however, has been revised and restated over the last hundred years, and new geological findings contradict his claims.

was inquisitive about the shape of the boulders and could not understand what could have got them there. His intense interest in these granite formations claims an involvement in geological and Romantic imagination, one that anticipates by a century the petrochemical and modern imagination involved by our interest in, observations of, interactions with, and anxieties about plastic. Darwin's imagination coalesced around affective-elemental states.

Surprise. Those large masses of rock were later called Darwin's Boulders: erratic and enigmatic, speculative and Romantic. Geologically, Darwin attributed the "erratic to ice rafting,"[3] but the enormity and certain kind of strangeness added to the wild beauty possessed his post-Lyell geological temper and imagination.

Bizarre. Granite has its own complicated formations, unstable and not simple in its petrological origin (as the "granite controversy" attests); "consisting of known materials yet combined in a secret manner, it is impossible to determine whether its origin is from fire or water. Extremely variable in the greatest simplicity, its mixture presents innumerable combinations."[4]

Becoming. "Granite genesis" has its competing explanations: "magmatic (granites are igneous rocks resulting from the crystallization of magma) and metamorphic (granites are the result of a dry or wet granitization process that transformed sialic sedimentary rocks into granite), because granites are the result of ultra-metamorphism involving melting (anatexis) of crustal rocks."[5] When H. H. Read observes that "there are granites and granites,"[6] he means an overwhelmingly variegated schematic emergence something that stares us as a prospect for a petri-becoming. It is imagination that informs the formations as we find metamorphic force, tectonic variety, chance,

---

3. Evenson, et al., "Enigmatic Boulder Trains," 5.

4. Sullivan, "Goethe, the Romantics and Early Geology," 346.

5. Chen and Grapes, *Granite Genesis*, 4: "geochemical and/or generic-alphabetical, i.e. S-, I-, M-, A-, and C-type granites (S = sedimentary source; I = igneous source; M = mantle source; A = anorogenic; C = charnockitic); calc-alkaline, alkaline, peralkaline, peraluminous, metaluminous granites; or are related to tectonic setting: 'orogenic' (oceanic and continental volcanic arc; continental collision), 'transitional' (postorogenic uplift/collapse), and 'anorogenic' (continental rifting, hot spot, mid-ocean ridge, oceanic island) granites."

6. Chen and Grapes, *Granite Genesis*, 4.

actional consequences, and unpredictability come into a complexity of emergence.

Goethe's "On Granite" (1784) speaks about the antiquity of Earth and a near unfathomable processual build up within Earth's crust—the Romantic and scientific imagination in a "becoming Earth." He observes that "composed of familiar materials, formed in mysterious ways, its origins are as little to be found in fire as they are in water. Extremely diverse in the greatest simplicity, its mixtures are compounded in numberless variety."[7] Granite has its own mix of quartz, feldspar, and mica, as Goethe in his essay "On Geology, Particularly the Bohemian" tries to analyze the mysterious trinity *Dreieinheit*. Heather Sullivan explains that "in his search for the principles of *Werden*, Goethe believes he has found the one inorganic material containing within it the promise of all other mineral structures. Much like his botanical studies where he seeks the form of the *Urpflanze* that is contained in each plant today, his geological exploration focuses primarily on the processes of development. Granite is the metaphor for, and physical embodiment of, the ageless laws of mineral formation. It is the solid form of becoming, *Werden*."[8] The solidity, interestingly, conceals a fluidity of "coming together," a kinesis in mineralization.

Conglomerate. Jason Groves argues that the word *Granitgeschiebe* indicates that "these 'granite boulders' are neither conceived in terms of static form nor in terms of the anteriority of ruins but rather in terms of an ongoing movement of ruination; rather than as a substance, granite is presented as a thing in motion."[9] Goethe's sensitivity to earth-formations foregrounds the principle of "incongruence" where the earth is left to transform and transit in a complex geoerotics. Reflecting on such "mineral actants," Groves observes further that "stones and rock formations present themselves, express themselves, transform themselves, let themselves be seen, produce themselves, spread themselves out, alter themselves, and conceal themselves [*zeigen sich, sprechen sich aus, verwandeln sich, lassen sich sehen, erzeugen sich, verbreiten sich, verändern sich, verbergen sich*]. In this drama of things, mineral agents

---

7. Bell, "On Granite," 861 (emphasis added).
8. Sullivan, "Goethe, the Romantics and Early Geology," 347.
9. Groves, "Goethe's Petrofiction," 100.

Figure 19. From *Plastiglomerates*, 2013. These found object artworks are the result of a scientific study by geologist Patricia Corcoran, oceanographer Charles Moore, and artist Kelly Jazvac. Photo credit: Jeff Elstone. Used by permission of the artist.

take humans as accusative objects: they direct our attention, they address us, they come together to make formations."[10]

The fascination and fetish, imagination and indignation today is with plastic. Our aggrieved and aggressive turn to plastic has brought us before a strange petri-kin to Darwin's Boulders—"plastiglomerate."

First discovered on Kamilo Beach on Hawaii's Big Island in 2006 by Charles J. Moore, who spotted "odd plastic-covered rock assemblages" along the trash-strewn stretch of sand, and named in an article in *GSA Today* by Moore, Patricia L. Corcoran, and Kelly Jazvac,[11] plastiglomerate (figure 19) is a mixture of plastic-intermateriality, comprising the surprising, bizarre, becoming, erratic, and aberrant. Corcoran, Moore, and Jazvac describe the new substance as an "indurated, multi-composite material made hard by agglutination of rock and molten plastic. This material is subdivided into an in situ type, in which plastic is adhered to rock out-

10. Groves, "Goethe's Petrofiction," 106.
11. Nuwer, "Future Fossils."

crops, and a clastic type, in which combinations of basalt, coral, shells, and local woody debris are cemented with grains of sand in a plastic matrix."[12] Plastiglomerates are widespread today, declaring nothing less than a "new epoch of Earth history."[13]

Plastic was imagination when it first burst into reality in the laboratory one hundred years ago and started affecting our lives and times. With time, plastic rolled out to become a reason to live by but having gone invisible and master-crafty it was in the realm of imagination again. Plastiglomerates, as well as such phenomena as the disemboweled "plastic" belly of the albatross, testify to the material's stealth and covert ingresses. Granite and plastiglomerate are "deeply embedded" in their own way—a poesis in petrology and the petrochemical, respectively. Plastiglomerate, like Darwin's Boulders and Goethe's granite, comes through as a "Romantic rock," speculative and spectacular, with a distinct geological aesthetic.[14] Like Goethe's "erratic" granite (*umherliegende Granitblöcke* and *Granitgeschiebe*) and Darwin's errant boulders, plastiglomerate, as an idiorhythmic, signals a fresh understanding in geopoetical thought and lithospheric imaginary.

If granite is the *ur*-stone of the Romantic age, plastiglomerate is the novice stone of the Anthropocene. One speaks of geological plasticity and the other a kind of plastic materialization, both attesting to an extraordinary geological becoming, vehemently pressing a fresh material-aesthetic into play. This newfound geo-reality emerged through an "intermingling of melted plastic, beach sediment, basaltic lava fragments, and organic debris."[15] If sedimentary or igneous rock speaks about the impact that a changing Earth has on its own formations, plastiglomerate leaves behind the traces of a virulently wasteful human behavior and what it has done to extinguish itself through an invention that ensures eventual ruin, while also ensuring that the components of our cataclysmic doom are visible. If granite grew out of a blend of quartz, feldspar and mica, building its own mineralization with certainty and mystery, the materials identified in the formation of the plastiglomerate speak of a larger story of civilization and its discontents and of a thinly-veiled

---

12. Corcoran, Moore, and Jazvac, "An Anthropogenic Marker Horizon in the Future Rock Record," 6.

13. See Castro, "Plastic Legacy."

14. See Heringman, *Romantic Rocks*; Lyell, *Principles of Geology*; Duncan, "On Charles Darwin."

15. Corcoran, Moore, and Jazvac, "An Anthropogenic Marker Horizon," 4.

dystopic controversy in the making—a "kind of junkyard Frankenstein."[16] As Corcoran, Moore, and Jazvac observe, while plastic melts beyond recognition in some plastiglomerate, in others it is recognizable as "netting/ropes, pellets, partial containers/packaging, lids, tubes/pipes, and confetti," as well as the "embrittled remains of intact products" in the form of containers and lids, ropes, nylon fishing lines, and parts of oyster spacer tubes.[17] Through plastiglomerates and other forms of plasti-petrification, "plastic geology" becomes our new speculative thought.

Here is how the artists Richard and Judith Selby Lang imagine this process unfolding:

> The history of the Earth can be read in the pages of geologic layers, built up sediments, igneous volcanic flows, rock stacked and folded until strata are formed. Each layer deciphered is an understanding of a past moment of natural history. In the year 2855 CE, a startling discovery was made that unfolded the mystery of what happened over 10,000 years ago, revealing clues of what happened in a time our popular press has come to call "The Age of Grease." It was a time when fossil fuels were sucked out of the earth, dug out of the earth to drive a civilization towards its demise. A layer of brilliantly colored substances was found sandwiched between layers of rock: substances that civilization called "plastic." The layer which is so distinctive that our geologists and archaeologists have come to call it a discontinuity, a term used to describe nonconforming conditions. Hence, "The Plasticene Discontinuity" is aptly named.[18]

We roll into plastic discontinuity through *altplastic*—the excess and excrescence that come in a variety of forms to trouble us both externally and internally. It is here that plastic inscribes its own time on geological time, affixing a historical-cultural text to it for posterity to decipher. The material-aesthetic of plasticization has started to build its affective imprints. Plastigomerate makes us read Earth through a historical and cultural narrative whose point of inception and emergence stay niched around the first decade of the twentieth century. This is the geological line, as part of Jason W. Moore's

---

16. Chen Jun, "Rocks Made of Plastic Found on Hawaiian Beach."

17. Corcoran, Moore, and Jazvac, "An Anthropogenic Marker Horizon," 6.

18. Lang and Selby Lang, artist statement for their 2004 exhibition at the Bay Model Visitor Center in Sausalito, "The Plasticene Discontinuity."

concept of *oikeios* ("a multi-layered dialectic, comprising flora and fauna" and "our planet's manifold geological and biospheric configurations, cycles, and movements"[19]) about territorializing Earth with clear indicators to separate a pre- and post-plastic geomorphology.

If the structure of rocks, be they "granite, serpentine, slate, sandstone, limestone, chalk and the rest," gives "rise to the psychic life of the land,"[20] as Ithell Colquhoun notes, what psychic life do idiorhythmic plastiglomerates produce? Material-aesthetically, what is the earth-spirit manifestation? Such earthly turn to plastic has changed geo-respiration. Paul Harris shows us how Pierre Jardin's "scrambled perceptual borders of stoned thinking" can be "aligned with the abundant new materialisms emerging in the geologic turn."[21] Geophany (in the sense of an epiphanous Earth) produces a new poesis and praxis of "listening": plastic makes Earth listen differently to a new anthropogenic production as much as humans are seen to touch Earth differently. The "bewitchment" of rationalism works differently in plastiglomerates where Earth is itself a collector, assimilator, and networker. Bewitchment is about fabricating the identity in a surreal geoscape. In a different kind of plastic art, objects disappear to surface underground through a separate material sense. Here, again, a new form of reality suspends human rationality, if not scientific explanation; it creates a fresh historical narrative. Plastic geophany brings a "break" in geomorphological continuity, articulating a "coming-present" in the process. Here one need not miss a "passive vitalism" that Claire Colebrook, working through Deleuze, explains as a "pre-individual plane of forces that does not act by a process of decision and self-maintenance but through chance encounters."[22] The plastiglomerates bring through such passive vitalism a geo-queering whereby the "plastic subject" loses the capacities to explain and order the world under its feet, not through sublimic unrepresentability but in an uncanny disclosure of the material assemblage. There is a "chance" in the "indifference" of altplastics, the *hasard objectif* (objective chance)—a sudden confluence point of contrasting objects. Geo-queering brings a randomness to the process—causality not without probability. Mark the nylon-inscribed plastiglomerate shown in figure 20:

---

19. Moore, *Capitalism in the Web of Life*, 46.
20. Colquhoun, *The Living Stones*, 57.
21. Harris, "Stoned Thinking," 146.
22. Colebrook, "Queer Vitalism," 77.

Figure 20. From *Plastiglomerates,* 2013. These found object artworks are the result of a scientific study by geologist Patricia Corcoran, oceanographer Charles Moore, and artist Kelly Jazvac. Photo credit: Jeff Elstone. Used by permission of the artist.

Plastic Earth—through the nylon on the face of the material—transcends the tradition of terraformation to sculpt out a new face for itself; it is an art form that has a conjoint intervention—human and the nonhuman—to it. Here is a displacement of agency that the invention of plastic a hundred years ago had never envisaged. The nylon-stamped rock announces a distant human hand. It is about how Earth separates from its inmates and from itself: altplastics remind Earth how it can surprise itself and be its own cause of wonder. As a geological and art object, the plastiglomerate straddles the inbetweenness involving the geomorphologically existent, the plastic re-objectified, and the expansion of the human-nonhuman affective arc of transmedial existence. The interesting dimension rests with the "unintentioned" expression of art; or should we allow ourselves to see the humanly unthought patterns of aesthetic geomorphology? This unintentioned expression is a new turn in plastic's contribution to geomorphological adjustments. It becomes an art built through the lack of control over plastic, helplessness before plastic and non-knowledge of plastic—the plastic arts, whose plasticity, unlike the aspect of plastic art-making that I discussed in chapter 3, is beyond human craft and cognizance. The impotentiality of plastic has factored in the "human" without knowing what processes inspired an unconcealment beneath the feet of anthropos; this is chance art, stealth art and also a deeply studious art.

Chance in the geological imagination is not without design and pattern, though it is often outside human reason. Plastiglomerates challenge the trajectory of human thinking: if plastic arts have seen the surprises of human imagination, this petri-art has surprised human imagination instead. This is not to deny how chance factors in during plastic art installations. But here chance has its own ways of performance where the apparent incommensurability between plastic and rocks is overcome into a new petro-aesthetic event—a dialogue with Earth we never thought was possible. Plastiglomerates exhibit a randomness and a disorder but the connection between the two is not a very tight one. The probabilistic entanglement of the nylon rope and the toothbrush is not completely independent of contingent and historical conditions that beset everyday life. It is mobilism and what Marcel Conche observes as "réduit l'être aux événements"[23] (reduce being to events)—that this particular system does not recognize the existence of substances: the world as a whole is made up of events, and only events. This event in geodesire is the "geometric locus of certain significant coincidences,"[24] where the meeting of objects bears out a reality and historicality and is presentist and futurist at the same time. Kathryn Yusoff observes that

> aesthetics might be considered as a transposition and transmutation of material and immaterial forces, which parallels life's iterations across the "field" without the prohibitions that govern the concerns of subject to be faithful to an image of itself. That is, aesthetics are not primarily about picturing or representing what life is, confirming its affiliations, and forms of self-witnessing, it is about experimenting with what life can or might be—painting the exteriority of inhuman intimacy and imperceptible forces that draw entities into being. In this sense, aesthetics is a space of actualization in the formation of subjectivity, but also in the counter-actualizations of the new, and beyond the temporal-spatial coordinates of the now.[25]

---

23. See Lejeune, *The Radical Use of Chance*, 43.
24. Barbiero, "The Object as Catalyst."
25. Yusoff, "Geologic Subjects."

Plastic rocks make for a "counter-actualization" of a "new" and a "now" that is processual and layered. Inhuman geo-art-forms revitalize the well-meaning and structured correspondence between the material and the aesthetic through disruptive temporalities and anthropogenesis that trigger a dialogue between the nonlocal and nontologic, between determination and speculation. Plastiglomerates are formed out of surplus—the over-spilling plastic—as much as plastic arts are constructed out of material-aesthetic superfluity. (In her *Coastline* series, artist Evelyn Rydz uses 327 fragments of "found rope washed ashore at different locations in North and South America"; "each frayed rope varies in form, having been individually shaped and relocated according to the ocean water's natural movement."[26] Rydz's art with nylon and the nylon-singed plastiglomerate are instances of the restive hyphen in our understanding of the material-aesthetic).

It is through "indifference" that altplastics touch our body differently. Objects from our physical space and use have found unusual and uncanny places to show up. The traces of objects of physical use make us see them as our other—the vulnerable other—with a feeling of being surrounded and overtaken by an object whose presence is far beyond the visible. Following Gijs van Oenen, I would like to see the plastiglomerates, in particular, in the process of "emancipation": it is the humans and the earth that conjuncturally require them to be "emancipated." The object and the subject are on an equiplane of actancy where the plastiglomerate is not simply a product of human action and dictat. They are formed out of the human and the earth to "emancipate" beyond their forming. Here lies the actorship with "recognizable ends and intentions." The event of "interactivity" lies in the plastic industry and the whole cultural institution of plastic, which came to accommodate us in a variety of ways. They habituated us into a passive set of norms as humans came to know the institutional manners of functioning with plastic—institutions appearing as an "externalization of human subjectivity."

How does one "appreciate" this plastiglomerate piece of earth-art? What kind of "participation" between the viewer and the object as art is required to make sense of the phenomenon? Scientific explanation aside, this interactivity does not put an imperative on viewers' involvement and their input. This participation does not discount the interactivity phase between the

---

26. See this gallery for the pictures: https://evelynrydz.com/section/454800-Unraveling-1000 -Years.html. Figure 7 in the gallery is by Rydz.

human and the nonhuman where affiliation to an institution (plastic) and its norms and forms of behavior (using a toothbrush or a nylon rope) have contributed to the making of the object at hand. But the interactivity ceases to be the reason when the plastiglomerate builds its own actional autonomy. Corcoran, Moore, and Jazvac have "interactively" engaged with the plastiglomerates to trigger an agency between the two parties. However, as van Oenen observes, the "emancipatory freedom of having to act only on norms one could subscribe to oneself turned into a virtual obligation to always do so, turning a blessing into a burden. Increasingly, we as human beings cannot bring ourselves to act on our own norms. Rather than admit defeat and give up on the interactively acquired capacities, we externalize, or outsource, them: we ask objects to exercise them, on our behalf."[27] Through "interpassivity," the object speaks, agencializes, and actionalizes. Under conditions of interpassivity, objects externalize themselves and "speak out" in a discourse beyond the logic of the human imaginary. Such "script loading" of an object directs it to us, and we become "aware of the fact that although we, as visitors, are still supposed to be present, and even interactive, our input is apparently no longer needed. The artwork has already taken care of it, leaving us redundant."[28]

The question of whether plastiglomerates have "insourced" our interactivity, rather than whether we "outsourced" it to them, brings us to wonder about a new form of emancipation for objects. Feeling responsible and feeling interpassive are different. Humans can claim responsibility for the plastiglomerates but that does not disallow a post-emancipatory condition. The plastiglomerate has insourced our subjectivity and rendered us emancipated: it is emancipated too. It has its own "design" in the sense of a pattern and intention. The plastiglomerate is not without us but within its formation and formability it has left us "without." Unlike plastic-art, which is interactivity, it is, within a specific context of understanding and through a distinct spin-off of the material-aesthetic, interpassive.

Plastiglomerate is irrational, perhaps, surrational. The earth surprises: it "ungrounds"; it is never the "solid ground" we thought it has been or was. Plastiglomerate has made it possible to see how rapidly the planet keeps transcending its status, topologies, and depths of geo-aesthetical expressions.

27. van Oenen, "Interpassive Agency."
28. van Oenen, "Interpassive Agency."

Plastic stirs; the earth stirs the plastic. Within the fluidity of a material-aesthetic, both address potentiation and possibilities, invent topologies and deliver freedom of expression. Earth is becoming-Earth—productivity preceding the product.[29] This transcendental geology leaves us to question the ground that has always been dynamical ungrounding. We are left wrestling with Jan Zalasiewicz's haunting question: What legacy will humans leave in the rocks?[30] Plastic petrifaction. As one of the major perturbation events, the culture of plastic will lie in the earth's crust waiting to be deciphered (the "plasticene discontinuity"). Overlaid, stratified and rarified, plastic projects the idea of an "impossible time," geological levels that are uncognizable, intuitive, and deep. As Zalasiewicz poignantly asks, "What might be the best fossil analogues for, say, a polystyrene cup?"[31]

Plastiglomerates metonymically declare that "the stratigraphic record has been set on an irreversible trajectory" with scalarity and temporality of a different order.[32] This process evokes a kind of imagination somewhat close to Lyell's emphasis on the mind that confronts an earth beyond its ken and comprehension. It is a geological imagination that surpasses a well-graded trajectory of human thinking. Lyell points out in his *Principles of Geology* (1830–1833) that "although we are mere sojourners on the surface of the planet, chained to a mere point in space, enduring but for a moment of time, the human mind is not only enabled to number worlds beyond the unassisted ken of mortal eye, but to trace the events of indefinite ages before the creation of our race, and is not even withheld from penetrating into the dark secrets of the ocean, or the interior of the solid globe."[33] Plastiglomerate sets off the "drama of geology" in a Lyellian way where imagination brings us before more earthly wonders. Is the plastiglomerate a "great stone book"?[34] What does it open up on and open onto? The interesting part of this development is in seeing how rocks that signaled "inhuman time" come to be "humanized" through material remnants of a particular species. If geological imagination is continuously under certain methodological ministrations to make the scalar and stratigraphic developments comprehensible to us, plas-

29. See Grant, *Philosophies of Nature after Shelling*, 169–170.
30. Zalasiewicz, *The Earth after Us*. See also Zalasiewicz, *The Planet in a Pebble*.
31. Zalasiewicz, *The Earth after Us*.
32. Buckland, "Inhabitants of the Same World."
33. Lyell, *Principles of Geology*, 102.
34. See Anstead, *The Great Stone Book of Nature*.

tiglomerate evokes a new imagination that projects its own post-anthropocenic potentials—a future that will only talk about the past.

Plastifossilization

Does plastic change the principle of causality in geomorphology? Traditionally, geomorphologists try to see a particular event processed out of antecedent circumstances; they "cause" the event "in the sense that there are certain empirical regularities" that declare "general and unexceptional connections."[35] Plastic brings unexceptionality to deterministic formations and often disrupts supposed linearity. Plasti-geology alters our notions of time contributing to a poetics of distance in which the Earth's crust becomes increasingly alien through the plastiturbation that transmogrifies geo-ontologies.

Seen in this light, plastfossilization complicates Jussi Parikka's concept of the Anthrobscene and its fossils of media waste. This plastifossilization is a relentless and inexorable process that recharacterizes an already unstable Earth—an Earth that is a "compilation machine, an assembly line," where "trash, construction debris, coal ash, dredged sediments, petroleum contamination, green lawns, decomposing bodies, and rock ballast not only alter the formation of soil but themselves form soil bodies, and in this respect are taxonomically indistinguishable from soil."[36] Seeping and sedimenting plastic keeps changing the soil's character and habit—a plasticization of the "soil-ego"—to create a deterritorialized Earth that is weighed down by irrevocable plastic sink. Parikka notes that "instead of the past cyclical formations returning in new guises, the future holds an alien visitor who finds traces of the earth molded by modern science. Future geology becomes a narration of history of human intelligence, meaning science and technology growing on top of nature, animals, and plants. It represents a novel geological periodization projected and then retrojected from the future back onto our world."[37] Parikka's "paleofutures" are, for me, "plastifutures." Plastic's agency changes our understanding of "relational materialism" and the non-humanization of geo-ethics.

---

35. Rhoads and Thorn, "Toward a Philosophy of Geomorphology," 123.
36. Parikka, *A Geology of Media*, 110.
37. Parikka, *A Geology of Media*, 118.

Plastic's existence has led us to post-environmentalism, post-humanism, and novel chemo-material initiations. Earth's deep time is increasingly being invaded by plastic time. Besides its non-plasticity and decay-resistant trajectory, plastic time includes forces that are global, capitalist, economic, and political. Plastic also changes soil-time, and the processes that make Earth's crust chemicalize in new ways involve temporality of a different order as soil organisms, including plants, now face a separate regime of existence and expiry. In these acts of re-materialization we are confronted with a new geo-entanglement. While older forms of fossilization generate interest in the earth-depths, plastifossilization has left us immobilized: deposition as depravity, discontinuity as despair.

We are born into plastic time. It is the endless flow of time, without decay, with a corresponding coordinate of time understood as advancement. This accelerates the contraction of biological time. Human time moves within the circumscription of plastic time; but plastic time, in its immovability and universality, impacts on human time and, hence, is responsible for a flow outside itself. This is non-degradable time for non-biodegradable materiality. Plastiglomerate proclaims no abrupt collapse of time and historical distance in understanding. It speaks, instead, of an evolutionary trajectory over the last hundred years that wedges the subject and the object. This is an anti-Romantic formulation that reverses subjectivity and in which object formation has its own precise scientific understanding and a clear heuristic discourse on the relational map with a transforming object. This is a complicated, though not incomprehensible, network of discovery, a process-point in the laboratory to a baffling consequence within the laboratory of the earth. Plastic geo-layering has its own genealogy that constructs a complex parallelism between the plastic materiality and biological plasticity in which running away from plastic is always already running into plastic. The plastic hardwaring of earth, therefore, develops its own vanishing and expiry moments, the points of sustainability and composite fractures—a kind of "variantology."[38]

If decay is the manifestation of vitalism and the forces of transformation, the decay of plastic (plastic time read as slow time) can be understood through what Mark Jackson qualifies as forms of interiority. He notes that plastic

---

38. Zielinski, *Deep Time of the Media.*

presents itself as an ethical concern for us precisely because it cannot be interiorized within the living horizons of the planetary bird, or similar earthly
life processes. It might be able to be interiorized within the life span of a min-
eralizing object as it will degrade eventually. But, if the ethical demand
comes to one as the attention to the problem of planetary life, whether human
or more-than-human, then the means by which the ethical demand is a question for thought must necessarily negotiate the fundamental and essential
principle for life of interiorization. And this process of interiorization, which
is the ground of ethical concern, is a process given to us quite outside human
intentionality. Indeed, it is the condition for human action and will. Thus,
the ethical privilege and demand come from outside us, before us, in the ontological process of life as capacity for interiority.[39]

This ethical concern leads to a new poetics of geomorphology that corresponds with "planetary interiority." The plasticity of plastic as artifact and
plastic as chthonic reveals the both-state existence in which the interiorization differs in ethico-cultural ways—the difference it makes to our living and
values of use and existence. We make plastic and the plastic unmakes us
in a kind of mutual recognition of transience, power, control, and impact.
The dynamic of planetary cohabitation comes to renarrate our "humic"
connect.[40]

Is plastifossilization mediated by the anthropos, the material habits and
consumerism and the global flow of trash and waste? This fossilization is a
process, a fluidity that constructs a reality of humanity's performance in living as well as its construction of an unknown future. If fossil fuels precede,
plastic fossils follow. This flow revises the whole question of temporality.
while organic fossils represented a transition from a prehuman past into the
human present, with plastic it is a present that prepares a future whose appropriation and accommodation is difficult to envisage. Plastifossilization is
not mere plasticene but post-plasticism—a state of *being with*—that is a co-
habitibility whose parameters will stretch the biology, physiology, and materiality of our existence and encircle, in the process, the human and the nonhuman. Life forms that can adapt will have an unprecedented life-structure,
both internally and externally, through post-plasticism. With plastic invasively

---

39. Jackson, "Plastic Islands and Processual Grounds," 216–217.
40. See Harrison, *Gardens* and *Juvenescence*.

scripting its own prints deep into Earth's crust—an addition to an "invasion ecology" as it were[41]—the anarchaeology (here in the sense of anarchic archaeology) of differing manifestations is evident. Plastic produced is plastic distributed; plastic transformed is plastic stored; plastic disused is plastic born.

Plastics, within a certain kind of material-aesthetic, have initiated the "concrete event" and the "sentential event." Calling plastiglomerate a mere concrete event (one fixed in time and capable of being explained more or less completely) can, as Carl G. Hempel notes in another context, be "self-defeating" in that "any particular event may be regarded as having infinitely many different aspects or characteristics, which cannot all be accounted for by a finite set, however large, of explanatory statements."[42] As a "sentential event," however, plastiglomerate encompasses historical and cultural explanations as much as geological and chemical forces. So existence is now plastic(ex)istence—our material ontology and nontology. Peace with plastic is impossible; inter-domainization is the reality as it refigures the subject-object divide in the argumentative field of Cartesian reason and also within what Michel Serres suggests is the overcoming of the human/world distinction. As plastic threatens to master the master-human, the fierce, narcissistic narrative informing our Cartesian settlement and topology in this world is challenged.

Plastic at large, in its non-laboratory avatar, is not limited to hylomorphic materialism. The turn to plastic has become a turn to "composite" plastic, dynamic multiform plastic, and "nonstructural" plastic in the sense of exceeding forms beyond human cognition and imagination. Plastiglomerates are part of "hyloenergeism," as Jeremy Skrzypek argues—plastics as "occurent continuants": "the term 'hyloenergeism' is, like 'hylomorphism,' a combination of two Greek words found in the texts of Aristotle. In the present case, the two words are 'hyle,' which is often translated as 'matter,' and 'energeia,' which is sometimes translated as 'activity'."[43] Hyloenergeism is the view that material objects are comprised of matter and activity, with activity playing the role of form. Plastics are both form and becoming-form: atelic and (a)causal. In a variety of denuded and denatured forms they script all these activities and mattering. Activation and actuality, potency and potentialization mark altplastic. Its manifestation and availability owe to a

41. Lockwood, Hoopes, and Marchetti, *Invasion Ecology*.

42. Kitts, "Geologic Time," 128.

43. Skrzypek, "From Potency to Act," 3.

"power" within—a deep structure—to which it contributes to maintain its identity as plastic in the teeth of a variety of other forces that are disintegrative and associative. Altplastics have built their own "transition" points of change and survival, dormancy and actualization, and difference and delay as units of impact. Changing properties of their own and altering the impact and ingress-points into nature and living bodies of human and the nonhuman other, plastic remnants have raised their activation levels, their structural power, and nuances of processuality.

Where do we identify the connectors between a plastic toy and the degraded ghost of its own form after unknown periods of forcible weathering? How does the toothbrush we use relationalize with the frayed and run-down toothbrush in the body of the plastiglomerate? Skrzypek points out that "a material object is not itself an activity or process, according to hyloenergeism; it is something composed of matter, which comes into existence when that matter is engaged in a certain activity or process. Understood mereologically, a material object is composed of both its matter and the activity or process that is occurring in that matter. And so hyloenergeism is not a 'pure process' ontology." Plastifutures, through synchronic mineralization, reveal altplastics that infuse each other through substructural changes and processual energy. The figurality of plastic as an event (in the sense of a formation) and occurrence (as a state of occurring) emphasize the "endurantist account of the persistence of material objects."[44] Plastic persists in its plasticity. Not merely committed to perdurantism, the "indifference" of plastic is connected to this persistence also.

Plastic brought a dramatic "difference" to our eco cultural lives; but in its invisibility and emergences in forms and shapes and avatars that dislodge our idea of planetary equilibrium, plastic has built an "indifference." On a significant material-aesthetic note, altplastics are persistent, percolative, and pugnacious as they find their poise and power through an indifference that is autonomous and authoritative. Plastic Earth lacks "shape" for it is ceaselessly performative and self-indulgent in response to the diverse contributions that altplastics never fail to make. Charles Scott observes that the word "indifference" stems from difference and *differre* and "also suggests possibility for change and movement toward unspecified differentiation. It suggests

---

44. Skrzypek, "From Potency to Act," 28.

possible determination without partiality and with neutrality toward the outcome—quite different from teleological intentions."[45] Plastic decides and determines its position and placement within an authorial indifference to its own plasticism—"the indetermination of indifference."[46] The irony of indifference is remarkable in that plastic is "determined" by humans—in its inception, proliferation, design, and application—but displaces itself from human agentiality through a purpose of its own. This purpose deactivates the source (the human, the industrial) and puts the consequences into activation with formations that are beautiful yet alienatory (plastiglomerates), tragic yet wondrous (beach plastics), differential but not without an indifferent life in motion. It is plastic indifference from commensurability (anthropogenic plastic) to incommensurability (the other side), from making a difference (plastic capital) to a difference of indifference.

Plastic Sublime

The techne, genius, and wonder of plastic are paradigms of what we may call "plastic ecology." It demands that we acknowledge the fettle and finesse of plastic as it configures our lives materially and aesthetically. The instrumentalist logic of plastic, its teleological agency, and its structuralist principles highlight the relationship and relationality of plastic and sense. Our life-world is informed by plastic sense. Plastic transvalues its presence, particularity, and possibility in that it relationalizes through a certain template of the given (the reality of living with plastic every day, the quotidian plastic); the waste to which plastic degradation leads (plastic discard); and ecologization of plastic into mutable matter (plastic future). Plastic sense is our new mathematics: (in)calculable, differential, equational, and relational. From a direct to a partial to a remote connection, plastic senses the coefficients of biotic and non-biotic existence. Physiological and geological, endocrinal and geomorphological, plastics are in different forms of mathematical predictions, statements, knowability, and formations. These different forms are another version of plastic mondialization wherein the discourse of plastic prevails and presides as much as the governmentality and technic-

45. Scott, *Living with Indifference*, 3.
46. Scott, *Living with Indifference*, 3.

ity around it. Plastic sense is about inhabiting technology such that the sense of plastic is our sense of existence. We are in the reign of plastic technology, which has redefined our eco-sensibilities to a point of changing our history and culture of sense. It is a combative and constellative existence between pre-plastic sense and the sense of plastic today. The world of sense manifests in plastic variation, in "becomings of desire" in which the underlying essence is plastic—a kind of pluralist ontology. Is this our new plastic logos? This plastic logos is the *legein* that is restless in its assemblage of a variety of forces and coordinates expressed in materialities, machine-desires, thousands of plateaus of economy, commerce, and bodies in motion. The plastic logocentrism has its own algorithms, diagrammatic and calculative manifestations and expressions, and aesthetics of growth-capitalism and dissemination: sensing plastic, plastic sense, sense *and* plastic all come together in a kind of plastic-pedagogy. Peter Haff's concept of the "technosphere"—a corollary and successor to the biosphere, comprising all of the technological objects manufactured and discarded by humans and the systems that enable this creation and destruction—has become, more specifically the plastisphere.[47]

The plastic objects swept up and "unearthed" reveal the threshold points of an anthropogenic understanding of Earth and its elementality and phenomenality. It is in the exceeding of the scope of human knowledge and systems of representation in the plastisphere that Amanda Boetzkes finds an "excess of the earth."[48] The "elemental" in artwork encourages a fresh sense-generation whereby nature comes to create its own forms of representation that challenge our limits of understanding about what makes an intelligible form. There is an "excess" that plastic Earth delivers whether through the land or the sea, overwhelming our categories of perception and possible world-formations. Earth othered through such an excess contradicts the instrumentalist and romantic view of the planet, adding a new value to our negotiations with it. Plastic formations, through land and sea, as demonstrated through Kelly Jazvac, Judith Selby Lang, and Richard Lang, can be called "elemental" because they cannot be precisely analyzed and made strictly analytical; also, they resist being understood as a totality. John Sallis insightfully observes that "nature sustains itself not only in proximity to the

47. Hörl, *General Ecology*.
48. Boetzkes, *The Ethics of Earth Art*, 3.

human world but also in a guise in which it exceeds this world."[49] Plastic nature exceeds our world-making, and this nature keeps re-turning itself in forms that we as humans can scarcely imagine and configure. It is a fresh, continually revised, sense-making. Plastic nature never returns to itself; it merely re-turns. It is strange to itself and becomes its own alterity. The plastiglomerates and the geo-sea profile come to reinforce our sheltering within the "elemental." Sallis points out that "fleeing to one's home as a storm approaches does not allow one to escape from the storm but only to shelter oneself from its force. Cultivating the field, fishing in the sea, and cutting wood in the forest do not open a path beyond the field, the sea, or the forest but rather constitute certain kinds of human comportment to these elementals in which one is encompassed."[50] Fleeing and staying away from plastic is "dwelling" in the plastic elemental.

As the material-aesthetic, the plastiglomerate is discrete and indiscrete object and object-sense. It is discrete in having plastic in it as a product of scientific investment and explorations; its indiscreteness results from an "objectification" (understood as a process) that invokes delightful horror, aesthetic dismay, and hysterico-sublime—science lends knowledge of plastic, but could not predict the plastiglomerate. It is "environmental" without being scientifically natured or generated. The "thin sublime," as Sandra Shapsay explains, emerges from sources that are physiological and transcendental and owes its allegiance to Edmund Burke and Kant.[51] She draws on Noël Carroll's essay on "being moved by nature," in which he argues for a pre-theoretical form of appreciating nature, such that

> when caught up in such experiences our attention is fixed on certain aspects of the natural expanse rather than others—the palpable force of the cascade, its height, the volume of water, the way it alters the surrounding atmosphere, etc. This does not require any special scientific knowledge. Perhaps it only requires being human, equipped with the senses we have, being small, and able to intuit the immense force, relative to creatures like us, of the roaring tons of water. . . . That is, we may be aroused emotionally by nature, and our arousal may be a function of our human nature in response to a natural expanse.[52]

49. Sallis, *Elemental Discourses*, 84. See also Metcalf, "The Elemental Sallis."
50. Sallis, *Elemental Discourses*, 95.
51. Shapsay, "Contemporary Environmental Aesthetics and the Neglect of the Sublime," 181.
52. Carroll, "On Being Moved by Nature," 172–173.

Plastiglomerate opens up an aspect of the sublime that is less "environment" and more "object processed." The "thick sublime" here calls on reason, rational reflection, and greater cognitive content: look what plastic has been doing in such beautifully horrific ways without our knowledge, at first imperceptively but now unavoidably, with such gradual lethal seepage and eventual toxicity! It is not a pain or exaltation that would need a Kant or Schopenhauer to explain and paradigmatize. This is not a strict dynamical sublime but has a certain intensity of emotion that derives from the fragility and precarity of humans' relation with nature. It brings more idea and intellection into our understanding of object and the object-processuality and is the process by which "scientific cognitivism itself invites a certain attention to the framing subject."[53] A sublime Earth-event through the plastiglomerate makes the aesthetic appreciator "take some notice of the relationships between the subject who frames and appreciates aesthetically and the environment to which she attends. Thus, in attending to a natural environment it seems perfectly legitimate, even in some cases required, to appreciate the fact both that the subject is part of that environment physically and provides the frame for her experience of the environment."[54]

What process enmeshed toothbrush or nylon rope with rock? What force and instance synergized plastic and rock, temperature gradient and flow, molecularity, seepage, and conjugation? The transcendence of plastiglomerate into an aesthetic object as much as a scientific wonder—affective-cognitive, imaginative-intellective—inspires a thick sublimity. It is a "sublime" that has found a "place." This is the "earth othered"; how does Earth "art" back? Here Earth art is a separate form of representation within the agency of Earth as it "appropriates" the human for paleo-artistic or petri-artistic representation. The plastiglomerate is a kind of Earth Plastic–art: potentials in speed, scope, and scale of geospherical expression. The interlocking of matter is unstable, unpredictable, and dynamic. If plastic clams its way into the digestive system of a fish and clanks out a space in the human body, it clings to rocks in exquisitely esoteric forms too. Earth delivers the "unknown"—plastiglomerate, in that sense, is an "unknown" experience.

The images of the plastiglomerate and the atlplastics speak of a defeated imagination because they emerge from "unrest." In line with Edgar Allan

---

53. Shapshay, "Contemporary Environmental Aesthetics," 195.
54. Shapshay, "Contemporary Environmental Aesthetics," 194.

Poe's "dark sublime," these images depict an alternative world of ecological violence, the uncanny, and "the unaccountability of experience itself."[55] Our guilt, compulsive agency, habit, and dread are inscribed, fossilized deep into the face of the rocks. Plastic's sublimity maladapts to a reality that, to a large extent, is of our own making; this keeps the "shudder" alive. The images punctumize our perverse imperfection within a planetary co-emergence; they don't encourage a supersensible substratum in a Kantian way so as to overwhelm us. The intensity of the impact is more at the level of seeing a dissipation of energy and an entropic disrupture within a system—the system of a plastic-habitual-compulsive living. The violence in the visual beauty of the rocks is in a Poe-like way anti-sublime, a transcendent moment that reveals our overwhelming decay and destruction. The rocks alienate us from ourselves, from the plastic-habituated species; they are a rupture in the formations of our subjectivities. As rude signifiers they become anti-Romantic to a point where we shudder to identify with our Frankenstein-other, despite seeing a clear reflection of our indulgences and irresponsibilities in their making. The sublime affect of plastiglomerates is dialectical: they make "wonder" a vexatious experience, collapsing present concerns with future-oriented dread, evoking an emotionally complex fear of beauty and raising the confusion that comes through the ascension of unreason.

At the same time, the "political" in such ecological sublime cannot miss discoursing our ongoing existence with a commentary on the future and by passing a judgment.[56] The "political" writes our history and our transpersonality with a nature that has its own "yet-to-be-seen-and-heard" apocalyptic soundings. Plastiglomerate speaks of a "plastic system" at work outside our cognition and cognizance, one that is at once alienating and, often, strikingly representable: rock-plastic hybrids, a flip-flopping plastic bag containing a trapped fish, a fishing net with a turtle caught inside. These ecological sights are potently political and figurally sublime, triggering what Matthew Taylor argues is an "ecophobia" through which fear "would remain suspended in the apprehension of this vulnerability, unable to vanquish its object because unable to regard it as either fully self or fully other."[57] Taylor goes on to argue that this fear recognizes the "self's integration into its envi-

---

55. Brickey, "Poe's Dark Sublime and the Supernatural."

56. See Kainulainen, "Saying Climate Change."

57. Taylor, "The Nature of Fear," 362.

ronment without the ability to overcome it . . . [and] rather than reinscrib-
ing a defensive dualism between one's self and one's context, such a fear would
be the inhabitation of a radically uncertain openness to the world. And if
this echoes the sublime, then it refuses sublimity's conventional reconsolida-
tion of the subject after its unsettling encounter with radical otherness."[58]
The plastic sublime, hence, lies in the disavowals of the "ordinal," the normal, the sane, the expected, and the commensurable. This makes for a distinct plastic sense.

The post-political dimension in such formations is hard to ignore as well.
Plastigomerate's affective potential lies in dwelling with disgust and dismay,
with beauty and artiness. This is an uncanny mix of values and emotions in
response to plastic narratives and several tales in its geological entanglements
(the plastic-stone hybrids, a structure yet exposed to structuration, and the
vexed intersections of economic, material, geophysical, cultural, behavioural
and habitual discourses and forces). Weiyi Chang notes that

> plastiglomerate delineates the edges of capitalism's contradictory logic, con-
> stituting a sculptural approach to the problem identified by [Charles] Moore
> when he argues that capitalism has effectively run out of new natures to con-
> struct and to appropriate. In so doing, plastiglomerate functions to embed en-
> vironmental concerns with broader and wide-ranging social and economic
> problems, rejecting the historical isolation of environmentalism as a niche po-
> litical interest group in favor of a view that entangles nature in debates
> around capital, labor, inequality, and materiality.[59]

Plastic as an "ideograph" has left nature deeply political—a plastic rock
within a dark ecological network connects with ruthless industrial disartic-
ulation and helpless consumption. Paradoxically, human survival through
plastic is human extinction by plastic, with troubling implications for justice
and equity, community sustenance, ecological sustainability and judgments.
The tribunal of plastic is here to stay.

If Kant privileges reason (*Vernuft*) over sensibility (*Sinnlichkeit*), confining
the emergence of sublime to reason and effectively shutting out nature, plastifossilization and geo-formations spring nature into an exclusive affective zone

---

58. Taylor, "The Nature of Fear," 362.
59. Chang, "After Nature and Culture," 110.

of the nonrational and asymmetrical negotiation with the nonhuman that is materialized out of human reckoning and reason. Within the anthropocenic art of disappearance, wherein hundreds of biotic forms are dislocated, thousands of species are rendered extinct, and coral reefs and fresh water bodies are becoming things of the past, plastiglomerates, through grotesque eco-engineering, are balancing the other scale with a slew of disturbing appearances: all are part of the planetary shift as the texture of terror undulates. With increasing authority, plastic, as fossil-energy consumption and eco-geo-assemblaging builds the "Nonlife" (in the words of Elizabeth Povinelli[60]). Plastic has made human habit subsume nature—the earth becoming an earthling of human-plastic habit. The planetary melt-down under plastic is irreversible—at this stage, it is both impossible and absurd to imagine a *sans*-plastic life—and the debt to a discovery is compensated through variegated returns. The plastiglomerate is one such anthropo-geo-contagion.

After ecological, nuclear, and chemical sublimities, we have the plastic sublime, which keeps narrating the scars of scalar upheavals, overwhelming history of consumption of a dominant species, social-economic subsumation, acculturation of habit and use, and a dismissive ruthlessness of planetary exclusivity that has all the miasmatic saturation of power and petrocapitalism. There is a plastic "shudder" that evokes a strangulation of reason and a surprise about another "reason" that brought actants into a synthesizing game to produce this anthropocenic art. It is the defiance of human reason—"an involuntary comportment" that sources one strand of the sublime—and "the liquidation of the I" as it perceives its "own limitedness and finitude" shaken and shuddered to the core of consciousness.[61] It is a present-future moment of "trembling."

The "speculative biologies" of Pinar Yoldas's 2014 *Ecosystem of Excess* installation embody this futuristic sublimic trembling. Yoldas envisions how the petrodigestive system in plastisphere birds will begin to break down plastic for digestion; the emergence of plastic balloon turtles that after "eating balloons for eons, evolve an elastomer back that inflate[s] and deflate[s]";[62] and plastisphere insects transformed by microplastics, which "are the main ingredients of the plastic ocean" and

---

60. Povinelli, *Geontologies*.
61. Adorno, *Aesthetic Theory*, 245.
62. Yoldas, "Plastisphere Lexicon." See also Mertens, *Pinar Yoldas*.

are the perfect sites for the oviposition of aquatic insects. According to a 2012 scientific study, there is a positive correlation between increased levels of microplastics in the neuston layer and the reproductive success of pelagic insects. As a consequence, the plastisphere opens a new chapter in the wondrous world of aquatic insects. With their state-of-the-art microplastics nests, their most delicate and nutritious eggs, and their wildly coloured bodies, plastisphere insects are key players in the Ecosystem of Excess.[63]

Such speculative earth life projects a post-plastic and post-natural world where being in plastic-toxic will be our new-found sublimity. This connects with Rachel Honnery's paintings of Plastosessile, which provide examples of the "first hybrid plastic and organic creatures. They are a nod to the organisms of the Paleozoic Era (the first Cambrian era, the development of multicellular organisms, around 500 million years ago). They are created to remind us that we can't always see the hyperobject; that occasionally it is too small and too big at the same time."[64] The material-aesthetic "senses" out and at the same time works within such speculative and realist sublimities—busy, becoming, and baffling in an "ecological vertigo."[65]

Plastipsychosis

The rock-body, in a distinct material-aesthetic, scripts the tale of a human body, too. If plastics can tear into the rockface to create plastiglomerate, they can sneak into our skin pores; perhaps more insidiously, the affective impact of imperceptible plastic can infiltrate the psyche, causing paralyzing, destructive despair. "What we touch, touches us,"[66] notes Tom Fisher: the material, the aesthetic, and their affordances. Fisher observes:

> Plastics cease to be pristine, and become evidently worn, in a particular way. They do not patinate; they gather dirt rather than "charm," and then may elicit particularly strong feelings of disgust. When they are no longer an acceptable element in humanized nature, they perhaps are doubly unnatural.

---

63. Yoldas, "Speculative Biologies," 60–61.
64. Honnery, "Absolute Kippleization and the Plastosystem," 68–69.
65. Honnery, "Absolute Kippleization and the Plastosystem," 12.
66. Fisher, "What We Touch, Touches Us."

They are not trustworthy because they seem to make an issue of the margins of our bodies, and the manner of their ageing draws our attention to their margins. Whether as a result of this or not, consumers seem particularly sensitive to the characteristics of plastics' surfaces and to know that, while they generally are impermeable, their surface often is porous. Plastics, therefore, may be physically "tacky" and engender fear they will pollute with invisible chemical components and absorb disorderly matter. This pollution seems to operate according to the principles of contagion and essence found in natural magic, principles that also allow plastics to be a vector for social or moral contagion.[67]

Materially tacky, contagious, and traumatic, plastic aestheticizes the predicament of two plastiphobes: one is Jody A. Roberts, an American Science Technology Society (STS) researcher who works at the intersections of emerging molecular sciences and public policy; the other is Mrs. Maya Chakrabarti, in the Indian novelist-poet Sumana Roy's plangently evocative story "Untouchability."[68] Both connect around the acerbic irony of seeing plastic as "untouchable." But how untouchable is plastic when, as Mrs. Chakrabarti screams, the "world's become plastic, space has become plastic, time has become plastic"?[69]

Speaking about his seriously ill daughter, Helena, right after she was born, Roberts was alarmed by the amount of plastic that stood dangling from her in the first few weeks of her life. Roberts recalls that "our (mostly) unmedicalized pregnancy and birth experiences were being replaced with a hyper-medicalized post-birth. When the doctor arrived, he immediately intubated Helena. The long plastic tube now connected her lungs to the outside world, ensuring that air would continue to pass into her lungs and preventing her body from deciding otherwise."[70] Through the proceeding weeks, "a variety of plastic tubes carried a variety of fluids from a variety of plastic bags offering pharmaceuticals, nourishment, and—just once—blood."[71] Initially distracted by the anxiety of seeing Helena recover, Roberts's plastiphobic nerves gradually started to flare and flap as he began to wonder about the

67. Fisher, "What We Touch, Touches Us," 30.
68. Roy, "Untouchability."
69. Roy, "Untouchability," 42.
70. Roberts, "Reflections of an Unrepentant Plastiphobe," 102.
71. Roberts, "Reflections of an Unrepentant Plastiphobe," 102.

chemicals entering Helena's "fragile new body along with the medications, and electrolytes, and eventually breast milk (lovingly and dutifully pumped through plastic flanges into plastic bottles and bags)."[72] The dependence and the psychological rupture that come immediately after underwrite the countertextuality of the material-aesthetic: its technoscientific amenability—the inevitability of tubes, syringes, spectacles, phones, and feeding bags—and inescapable physiological intrusion.

In a similar way, Mrs. Chakrabarti, in throes of trepidation, refuses the milk man because the milk came in plastic packets, chooses not to clean the plastic dump in her apartment with rubber gloves, consternates over making phone calls for phones are made of plastic, refrains from using plastic buckets and mugs in the bathroom, stops wearing her spectacles as they, too, contain plastic, and is revolted at the prospect of touching her hair with a plastic comb. A plastiphobic psychosis riddles the everyday of Mrs. Chakrabarti and Roberts as their lives are interwoven with a material whose effects are disruptive and indeterminate. Roberts confesses, "When I look at a teabag steeping in a cup of hot water, I can't help but wonder what the bag is made of and whether or not it is breaking down in the presence of near boiling water. When I plug in the humidifier in the bedroom to combat the dry heat of our radiators, I wonder what might be accompanying the steam as the water exits its polycarbonate container. I start to wonder where all of the finely ground pieces that once constituted my shoe sole end up after they have worn off."[73]

Living away from plastic is living with the anxiety of how not to use plastic. Mrs. Chakrabarti is eventually admitted to the hospital because of under-nutrition and life-debilitating fragility: "She'd stopped eating almost completely—there was nothing, almost nothing, that didn't come inside a plastic container or packet anymore—rice and dal, oil and spices, vegetables and meat."[74] The tremble and shudder around the visible and imperceptible plastic continue. Ironically and yet pragmatically, Roberts admits that

> for better or worse, our lives are dominated, and in a tangible and real way made possible, by the very plastic that I've tried so hard to avoid and protect

---

72. Roberts, "Reflections of an Unrepentant Plastiphobe," 102.
73. Roberts, "Reflections of an Unrepentant Plastiphobe," 104.
74. Roy, "Untouchability," 49.

us from in our daily lives. It's the bottle that holds milk; the tubes and flanges that make pumping possible; the bladder bag that holds the milk to be delivered; the tube that connects the bag to Helena; and of course the g-tube crossing the space between the outside world and the inside of Helena's stomach (or, previously, the NG-tube that once dangled down her throat and into her stomach).[75]

Roberts understands the inevitability of plastic. Mrs. Maya Chakrabarti knows staying away from plastic is "maya" (illusion). We are deeply wedged in the plastihabitat; touching the "untouchable" is a form of transcendence in which extinguishment, in a searing irony, lies in the touch and the nontouch. The touch has been so profoundly unavoidable that it has conquered untouchability.

Plastics have been infiltrating the human body for the last eighty years through baby bottles, cooking pans, plastic films, wraps, packaging, preservative cans—and through overlapping industrial, domestic, and professional spaces. Like plastic-altered rock and earth, the plastic-altered body has emerged as the material becomes the axis and the "toxic allure" on which we have come to depend.[76] Questions remain as to how somatically transformed we might become as we negotiate Earth's plastic turn. The biological realities of the body and the historical realities of plastic come into vexed contact with values, adaptation, submission, and scales of toxicity and malleability. A vexatious poetics of imperceptibility attends plastic with an intangible affect whose manifestations are outstandingly different and miasmatically varied. Plastic and plasticizers imperceptibly break through immuno-physiological systems in humans and nonhuman animals, evoking bodily insurgency. The material body aestheticizes with a difference as we encounter a burgeoning world of "phthalates syndrome" through our absorption of EDCs (endrocrine disruptor chemicals, a term coined by biologist Ana Soto). As detailed by Max Liboiron, "Various plasticizers have been correlated with infertility, recurrent miscarriages, feminization of male foetuses, early-onset puberty, obesity, diabetes, reduced brain development, cancer and neurological disorders such as early onset senility in adults and

75. Roberts, "Reflections of an Unrepentant Plastiphobe," 112.
76. See Chen, "Toxic Animacies, Inanimate Affections."

reduced brain development in children."[77] In nonhuman animals, EDC exposure has been connected to reproductive and developmental abnormalities. Alligators in Lake Apopka in Florida have been "feminized" with "males having shortened penises and low levels of testerone, while the females have excessive levels of estrogens." Sex reversals and "skewed sex ratios" have marked several species of fish. Confusion remains as to the amount of plastic seeping into Earth's crust needed to change rock formations; confusion also remains as to the dose-intensity of EDC that might start sexual dysmorphology.[78] Heather Davis notes that "in addition to the outright transformation of the normative signs of gender, exposure to plastic chemicals has also been shown to affect behavior."[79]

The complexity of these developments is not manifested through singular events; the disruptions and derangements come in a web, in nonspecific zones of impact, in a kind of "cocktail effect." The body starts to behave outside itself, non-habitually, through temporal distension wherein organs and organicity are pushed to a point of revolt and antagonism. Like geo-queering as evidenced through altplastics, does the body "queer" up? Do the plasticizers bring abnormalities in human bodies resulting in transformation of gene expression with physiological disruptions and imbalances that closely connect with the interests of transgenderism and bisexualism? This I call "slow sexuality," a teleology that has its own modes and temporality of progression—plasticized sexuality. The "miasmatic geography of plasticizers" (as also phthalates and BPA) is overwhelming: the human body cannot avoid plasticizers anymore. Plastic's biochemical performativity is about a "becoming-plastic" body that builds its own discourses of strangeness and unfamiliarity with itself and the world. This is the "meshed" body where "boundaries between self and other, biological and chemical, living and inert," as Katie Schaag argues, dissolve to change "an organism over time, mutating it towards otherness from itself."[80] The hyphenated space in the rock-living body event highlights an experience that is held in knots at unexpected places addressing unexpected manifestations: porosity of both and

77. Liboiron, "Plasticizers."
78. Hood, "Are EDCs Blurring Issues of Gender?"
79. Davis, "Toxic Progeny," 238.
80. Schaag, "Plastiglomerates, Microplastics, Nanoplastics."

in both and the disseminative and dissipative character of plastic meet in exchanges of deficit, excess, neutering, mutation, and abnormalities. Plastic phylum, material-aesthetically, sets a queer future for both the bodies that articulates disruption as a continuity of geo-bio-life: not the end of a future but a queering of a future disembeds how we think and practice reproductivity, build and sustain progeny, the politics of generation-building, species orientation and extinction. Life is plastic continuous, and this can create a sense of curiosity and perhaps, even "pleasure" at the "queer plasticization of our interiority"[81]—plastic acclimation.

As with the rock so with the body and so with the psyche: altplastics in their lithospheric seep and aqueous spread create a psychosis that suspends our sense of oneness with the growing aberrant world; they disable our sense of engagement too. Psychoanalyst Harold Searles argues,

> We equate the idealized world of our irretrievably lost childhood with a non-polluted environment. We tend erroneously to assume that nothing can be done about the pollution of the present-day environment because of our deeper-lying despair at knowing that we cannot recapture the world of our childhood and at sensing, moreover, that we are retrospectively idealizing the deprived and otherwise painful aspects of it. The pollution serves to maintain an illusion in us that an unspoiled, ideal childhood is still there, still obtainable, could we but bestir ourselves and clear away what spoils and obscures its purity. In this sense, pollutants unconsciously represent remnants of the past to which we are clinging, transference-distortions which permeate our present environment, shielding us from feeling the poignancy of past losses, but by the same token barring us from living in full current reality.[82]

We see an "intimate kinship" with this growing deviancy around a structural kinship that imperceptibly decides our biological fate—an extinction that is different from death, a thanatic spectrality removed from mere mortality. The egoistic understanding of the life that we think comprises our world is interrupted by plastic; the ego-boundaries are lost. The nonhuman reveals through us and in its perpetual becoming makes us regress ontogenetically and phylogenetically. The very sight of a plastic dump, the Pacific

81. Schaag, "Plastiglomerates, Microplastics, Nanoplastics."
82. Searles, "Unconscious Processes in Relation to the Environmental Crisis," 367.

gyre, the plastiglomerates and their intrusive counterparts in the sea beach and the sea, have psychoanalytically left us in "chaotic disruption" in relation to the world and the situations in which we exist and live. These are psycho-neurotic expressions of a rebellious id unwilling to accept the increasing reality of a garbage-tomb earth.

However, plastic affect confirms a greater unity in the sense of a bondage to or embeddedness in a planet permanently in the grasp of a nonhuman on which our dependence, both conscious and unconscious, is helplessly absolute. Plastics have shrunk our space on Earth as they have controversially increased their own. Is Earth "repressed" by plastics, sczhizophrenicized by plastic? Altplastics conjure a "turbulence" and create difficulty in identifying an "analytic" because the repression, rupture, and retreat are overwhelming for the individual and eco-social psyche. They surface in a zone of the "unthinkable" and the "inarticulable"—the psyche stressed into inarticulation by critical catastrophism. Plastic-embeddedness and plastic-becoming create a reciprocity that is built on the axis of mutual damage where the silence of earthly consumption, the omnipotence of the plastic Earth, and the plastic transference from human addiction to geo-repression develop their own signatures. Altplastics leave us with an unhomely Earth.

How much of Earth can we claim and how much can we leave to Earth itself, by living through dissociation, disrupture, and an affectivity that is essential to its sustenance? Altplastics build affective disconnection with the anthropogenic ego: here, projecting the possible extinction of future generations, plastic nature in its increasing hegemony and independence rivals the phallic control over nature by challenging the ego's authority and dominance. This brings the melancholy through the unhomely; and a relentless uncanny-ization deauthorizes the anthropo-techno-master ego through ceaseless plasticization as it unconsciously undergoes alienation from the future inhabitants (across species) of Earth and becomes the point of apathy for generations to come. Through regression and a helplessness against the inscrutable, the ego starts to feel displaced by the ever-increasing versions of altplastic. It is the strangeness as psychical resonance that struggles with changing forms of availability of the plastic earth-receptacle, the earth-womb, what John Keene has called the "toilet-mother."[83] Seeing the plastic earth is castration

---

83. Dodds, "Minding the Ecological Body."

anxiety as humans—plastic consumers—ignore constraints and compunction in the face of incessant mother-dominance. A conscious defense of the ego against plastic earth becomes sadistic when the geotrauma tips over an apocalyptic threat of self-extinction. This is just not mere death by plastic but a quiet, though certain, losing of life to a plenum of death forms, death-drives. Plastic life is about the eroticization of death. Altplastics are disturbing products of drive theory—the over-consumptive anthropogenic fixations to exploit the Other. Our negotiations with nature correspond with our growing intrapsychic frameworks where building relations is mostly about compromises in power, "defensive splitting"[84] that accompanies stress, dependence under dominance, and dominance through exploitation—the drive motor leading to death-drives. Plastic has evolved to become the "drive," the aggression and libidinous discharge between human and Earth. The complicated death-drive involves an Earth that is nearly overrun by plastic and species extinction and our pleasure-seeking plastiphilia: plastic civilization and its (dis)contents.

Plastiglomerates build a nonhuman subjectivity, instilling a "schizoid fear." Thinking through Harry Guntrip's schizoid phenomena, I see in such plastic invasion—through plastiglomerates, the plastic gyre, and plastic ingestion—the dread of being "smothered," "tied," "possessed," "dominated," and "absorbed."[85] Plastics have turned up our psychosis with a constructive commitment to change; also, it is staggered by a failure to resist effectively. A sense of failing sets in as the "bad" object relations impact on the ego in "smothering passivity," and when the impact tends to over-develop, it leads "either to schizophrenic terror of disintegration under violent persecution, or depressive paralysis under merciless accusation and pathological guilt."[86] Plastiglomerates metonymically propose the "futur ante'rieur"[87] that comes through a syncopated collusion between past and future while our "plastic subject" (in the sense of plastic-using subject) gets thrown into a state of becoming. The surprises that plastic reveals to humans—whether through the plastiglomerate or the objects that artists have displayed—interrupt the linearity and regularity of our existence with something

---

84. See Bigda-Peyton, "When Drives Are Dangerous," 261.
85. Guntrip, *Schizoid Phenomena, Object Relations and the Self*, 34.
86. Guntrip, *Schizoid Phenomena*, 84.
87. Gersdorff, "Schizoid Phenomena, Object Relations and the Self."

"new." Our geo-truths are punctured; the order of understanding is pierced; we become a part of the geo-trauma. Through what I call "plastizoid phenomena," rock-plastics bring out the affect of "plastic wounding." It has made anxiety a free-floating entity. What was envisaged as an elegant and convenient order when ushered into the capitalocenic space a century ago speaks of a disorder, a distress, and a deep wound, which in an affective turn has started to construct a fresh existential controversy.

Altplastics leave us with anticipatory mourning as every rock and rock-like material with plastic stains and striations—not to forget the ones that are growing imperceptibly and silently—leave us with a stress disorder. A certain kind of biophobia engrips us as an under-sea and above-ground retinue of plastic deaths of other species opens us to our "lack" in plastic exhaustion and depression. The mourning is our present continuous. And this mourning without finitude is a complex system and a changing pattern of nonlinear relationships wherein plastic cuts into Earth and Earth cuts into its inhabitants. The plastic affect has initiated a plastic ego whose embeddedness in the nonhuman environment contributes to its individuation. The individuation stems from plastic as a "transitional object" whereby it becomes a part of the formation of the plastic ego and builds affective and analytical links with the object as a part of the nonhuman world. The plastic ego, however, is struck by regression—interruption with the nonhuman—as it starts to live off and feed on the event of a rupture-as-negotiation with a nonhuman material whose differentialities and manifestations are enormous. Plastic, as nonhuman nature, extends the boundaries of the ego by making the unconscious hold the "rupture" as repressed. The foregrounding comes from a sudden but sustained exposure to a chaotic nonhuman nature. The ontology of the plastic ego builds on these regressions and attachments to object relations; it is recognized through the recognition of the other. The affectivity produces its own symbolic value.

Environmental melancholia springs mostly from the denial and disavowal of plastic as sublime toxicity.[88] The "being there" of plastic is also the agony of never being without. It's a failure and, hence, a regression that antagonizes rationality and economic reason. Jody Roberts and Mrs. Chakrabarti in Roy's story become irrational through a breakdown of rational egoistic

---

88. Lertzman, *Environmental Melancholia.*

understanding of plastic notoriety and noisiness—what Rennee Lertzman calls "an arrested, inchoate form of mourning."[89] It's not the "myth of apathy" among humans. Lertzman argues it is care that is present in surplus but "fraught with psychosocial complications and constraints,"[90] resulting in hesitation, doubt, and withdrawal. Can such care to avoid plastic and an apathetic vehemence to stay away from whatever is plastic help Mrs. Chakrabarti and Roberts to survive and sustain? Has protecting others, human and the nonhuman, become dependent on and obligated to plastic use? This is the mourning over plastic paralysis, where guilt, projection, dependence, strength, hesitancy, despair come in a convulated psychosis to argue the protection of the human as against the protection of Earth.

Lertzman and Theodore Roszak see "fragile psychological complexities" in humans that demand to be addressed sensitively and not be bludgeoned into a movement separating the "sympathetic to plastic-harm" from the "apathetic."[91] In line with Roswak's arguments, we may see the altplastic as a "psycho-social bundle" that overcomes simplistic guilt to reignite love for the earth as "care" for the earth. A dialogue with "inevitable" plastic and moderately and sensitively instrumentalized plastic comes into play. This can be a different turn to plastic. The plastic in the plastiglomerates, thus, material-aesthetically, has a bigger "affective" story to narrate. We embraced plastic (for its convenience, its strength, its short-term affordability, its permanence, even its beauty, among other things), and plastic in turn achieved a kind of "virality"[92] that is at once representable and involuntary in processes of encounter—folded, embedded, and affective. Plastic is our reformulated sczhoid in psyche, economy, history, anthropology, reflexivity, wonder, polymeric enquiry, and sociocultural unmoorings.

---

89. Lertzman, *Environmental Melancholia*, xiii.

90. Lertzman, *Environmental Melancholia*, 126.

91. Fisher, "Going Deep." Fisher explains that Lertzman's "aim is to cultivate a 'politics of environmental advocacy attuned to issues of despair, paralysis, anxiety and related emotions' rather than a 'politics of guilt or "feel good" steps to save the planet.' Environmental efforts that are 'quick to direct people into action' do nothing to address the 'powerful and inchoate affective investments, memories, desires, and losses' that are informing our environmental responses. She therefore wishes to replace superficial and simple behavior-change strategies with a depth-oriented and complex approach that works with the affective and unconscious dimensions 'underlying particular attitudes, beliefs or opinions.'" 222.

92. Sampson, *Virality*, 5.

The plastic turn, whether through plastifossilization or plastipsychosis or plastic sublime, is, to a large extent, agnotological. Plastics encourage agnotology as much in the earth as in a living body: an absence, hence, overlooked, hence, forgotten, hence, ignorant. It is a "steadily retreating frontier" brought about by "neglect, forgetfulness, myopia, extinction, secrecy or suppression."[93] Plastic makes the "missing" possible and the "missed" possible as well. Plastic agnotology brings its own versions of surprises as altplastics distinctly demonstrate. Agnoses in the plastisphere can be categorized as "cognitively inaccessible," epistemologically inert consequences of "specified ignorance" and the products of unknowability caused through "causal indeterminacy or contingencies of 'choice, chance or chaos.'"[94] The scale ("the components and systems" in which knowledge circulates) and granularity (knowledge and understanding as classified and censored and finely grained through analysis) of such plastic agnotology are patterned and pattern-less, structured and structure-less. This has made plastic affect a tantalizing experience. Knowing about plastic in heavy granularity of knowledge comes to be challenged through a helplessness before consumerist and capitalist hegemony and exploitation (human chauvinism).[95] Often this agnotology springs from unintelligibility, the inability to scale and chrometricize the plastic presence and deposition and a lack of care of the self and oneself through the ignoring of the other. The "plastic controversy" is, often, a reality intentionally ignored to perpetuate the conditions of present precarity—"malagnogenesis"[96]—where plastic-harm is strategically normalized and is accepted as an event within a humanly imaginable scale alone. The plastic accumulating outside

---

93. Proctor and Schiebinger, *Agnotology*, 3. Iain Boal came up with the term *agnotology* in the spring of 1992. "Ignorance in Greek really has two forms: *agnoia*, meaning 'want of perception or knowledge,' and *agnosia*, meaning a state of ignorance or not knowing, both from *gnosis* (with a long o or omega) meaning 'knowledge,' with the privative (negating) *a* prefix" (27). Ignorance of what plastic can do and is doing and the lack of perception that comes from indifference prepare a complicated negotiation between the material and the aesthetic.

94. Croissant, "Agnotology," 20.

95. See Croissant, "Agnotology"; Rappert, "Present Absences."

96. Martyn and Bosman, "Post-Truth or Agnogenesis?" Martyn and Bosman explain that "malagnogenesis (1) arises amongst unequal power dynamics, solidifying the status of certain groups as at-risk; (2) is often the result of agnogenesis; (3) leverages risk as a knowledge practice to normalize risk centrality; and (4) is *ipso facto* a bad thing. In other words, malagnogenesis takes root and is most harmful where an existing value gap predetermines marginalization and social exclusion. A cumbersome term, perhaps, but its virtue is that it draws a clear distinction between certain necessary, socially beneficial lies, and the malicious lies that are ever more in evidence today," 956.

our haptic-optic formulas of knowing the present is considered unintelligible, invisible, and consequently discardable. The agnotological plasticene is lived through quanta of unrealized reality. Whether it be known or unknown, visible or invisible, living the plastic event has impacted our "being-nature."[97]

97. Hamilton, *Defiant Earth*, 7.

Turn on . . .

Plastic is inevitable, the immovable ordinary ("Vinyl is as natural as lichen," as Christopher Dewdney states[1]). It burst into our cultural-material life with a libidinous energy only to establish alienation from nature, complicating our connections with land and sea. This is the moment that makes our idea of love confounding in the times of plastic. In complexity built through a variety of forces—economic, labour-intensive, garbological, ecocidic—love for plastic translates into a necessity to love avoiding plastic. Plastic affect works on a "disadjustment" (in the words of Bernard Stiegler[2]), a disorientation, with the biotic/non-biotic world, that builds more on mutual becoming than permanent split. The technicity through plastic is the "failing": the failure of nature and the human to hold on to any essence, a preformative being. Plastic today, rather, has become the "prostheticity," inculcating a separate vein of creativity and techne that keeps accentuating the

---

1. Dewdney, *The Secular Grail.*
2. See Lemmens, "This System Does Not Produce Pleasure Anymore."

"failing." Plasticity through plastic is supplementarity—it gives failing a new poetics and meaning by making humans surrender to economy, culture, forces of globalization, and inexorable precarity. Stiegler observes that "general organology defines the rules for analyzing, thinking, and prescribing human facts at three parallel but indissociable levels: the psychosomatic, which is the endosomatic level, the artifactual, which is the exosomatic level, and the social, which is the organizational level. It is an analysis of the relations between organic organs, technical organs, and social organizations. As it is always possible for the arrangements between these psychosomatic and artifactual organs to become toxic and destructive for the organic organs, general organology is a pharmacology."[3]

Plastic technology or plastic as technology is the unavoidable connection that ensures the cosmos of our everyday in society, economy, and culture and sunders us from the cosmos we chose for plastic to inhabit. It is a technological bios in serious irreversibility where living is not without the exhaustion of what constitutes and supports living. Plastic pharmacology is our new reality. And, in the Neganthropocenic life-world, plastic pharmacology is entropic in that the being and becoming are embedded in a discourse of present-future—toxicized by plastic and technological transcendence through plastic. But here comes a love that deindividuates us, atomizes us against a transpersonal existence with our surroundings, our nonhuman life, outside the eros of a plastic ego. The vector of desire has changed but the change comes on an axis that is difficult to enunciate—from desire for plastic to a desire against plastic without a helpless submission to the love for plastic. Plastic individuation is much more complex than we think it is.[4] At the core of such a complex love is "unfulfillment"—the eros to crave for more plastic riding on an unfulfilled desire and the unfulfillment arising out of an impossibility to eradicate plastic from one's life. Individuation is constructed through such an event of "to come" (*le devenir*) where a termination of both kinds of desire is near impossible, thereby making "unfulfillment" thrive as "desire continuous." Plastic affect is living with an unfulfillment. Plastiphilia is undoubtedly a deeply turbulent phenomenon.

Plastic enjoins and separates at the same time, and in a dyschronic mode of the eco-eco (economy and ecology) the desire for plastic changes its vector

---

3. Stiegler, "General Ecology, Economy, and Organology," 130.
4. See Hughes, "Bernard Stiegler, Philosophical Amateur."

from the aspiring to the alarming, from want to rejection. Plastic promotes "interconnectedness," and it also cannot avoid promoting "separation." Frédéric Neyrat sees "ecology of separation" as a contested principle where he believes that "interconnection must leave room for separation and must metabolize, symbolize, recognize it, if it is to avoid falling into the confusion resulting from the abolition of differences."[5] The material-aesthetic builds through transcorporeality, interobjectivity, and also a separation that contributes to a different kind of imagination and rationality. This separation, in fact, helps us to understand our interconnection and intermeshed state of being in a better way. Neyrat notes that for an ecology of separation to be political, one needs to take into consideration "the dangers which may threaten us, to distinguish between that which humans may construct and that which cannot or should not be constructed, to know in what ways it is still possible to use the words 'nature' and 'environment,' to enable the ecosystems to be resilient and to endure the disasters of the Anthropocene, ecology must leave space for separation."[6]

Neyrat opposes relentless constructivism that has a hegemonic hold over nature and ecology today. The ecology of plastic is constructivist but not without an imperative to judge it from a distance and a detachment that do not attenuate the politics of its status and being in the world. The complexity of the material-aesthetic is invested in inter-materiality and a mediation coming from a split (*clivage*): relating with plastic without an a priori investment and involvement. Living on the split has brought a "plastic nature" against a constructivist one—a nature technologized with a nature that has built its own technitis and oriented its own technophilia like the plastigomerate. This split takes constructivism to the other end of ecology where nature separates itself from the human-engineered nature, and, plastic mediates to accentuate the separation. The needle stops at the question on what eventually effectuates geo-constructivism. Plastic has united the world, has introduced interspeciesism, and has plasticized a community of thinking on a certain line of inevitable production and degeneration; plastic, again, has built a separate strain of eco-logics about which we can only fear in unknowability, speculate and imagine, ignore and withdraw in helplessness and arrogance.[7]

---

5. Neyrat, "Elements for an Ecology of Separation," 101.
6. Neyrat, "Elements for an Ecology of Separation," 102.
7. For more on "yet to be born" nature, see my "Globing the Earth."

The evolution of Earth is ceaseless, and its relation to technology has been diverse and increasingly intense. Earth has been terraforming and territorializing itself through an enormous complication in time and materiality. It is in the twentieth century that plastic has come to bear its input into geo-construction. No wonder we survive on the folds of a technical and "natural" nature that demands a separate ethics and sensitivity of understanding. Neyarat's arguments for geo-engineering as turning into "anaturalist nature" have their own force. "Climate engineering," writes Neyrat, "considers itself as ready to save the planet—even if we have to pay for it by way of some collateral damage, such as with periods of severe drought in equatorial Africa and certain parts of India."[8] If terraformation and geo-constructivism are plastic-driven, how can we come out of a territorialized Earth and the increasing intricacies of geo-material-abiotic relationships? Clearly, plastic affect has thrown a dialectical nature to live in and live for; it is not mourning over plastic for plastic has become a life within a life, a nature within a nature, a death outside our conservative understanding of mortality. Earth-life and earth-formations are not without a plastic immanence that has come to seriously attenuate the divide between the natural and the non-natural. Plastic materialism is changing the historical-geological arguments that we make with Earth and its planetarities (singularities, for instance). Earth is increasingly "eccentric";[9] it knows its ways to settle with plastic and allows constructivists to resettle it within a plastic-free "campaign" and technological intervention. Following Neyarat, we may argue that the plastic unconstructable is not what "eludes all construction" but is the unassailable and the essentially terrestrial. Altplastics and whatever comes undocumented and unprojected after them are "the inaccessible transcendental of production" that Neyarat calls the "transcendental dark side or counter-lining."[10] The transcendental is not strictly Kantian but a constructible that, by virtue of its inaccessibility and ungraspability, is unconstructable. The unconstructable, writes Neyarat, "is not an event that took place just once, at the origin of the world and of things; it has been present throughout the entire history of the earth, connecting its past with its most distant future: the transcendental dark side or counter-lining is the other side of the concrete history of the earth." Working outside

---

8. Neyrat, *The Unconstructable Earth*, 32.
9. Neyrat, "Eccentric Earth."
10. Neyrat, "Eccentric Earth," 5.

the geo-constructivist earth-object and neo-organicist earth body, there is the "earth-withdrawn," nonobjective, and "unthinged" event that connects the past with a non-projectible future. Something is there that does not love being graphed and grasped. The "eccentric" Earth makes its own claims as the eccentric living bodies declare eccentricities that are not subjectified abnormalities but the "counter-lining." Plastiglomerates are formed under conditions of human use of plastic; the plastic unconstructable will emerge under "the condition of the earth."[11] This makes for a fresh material-aesthetic.

As I finish writing this book, the world stands tormented under the Covid-19 virus. Earth is not a virus that would settle into a cohabitative peace with the discovery of a vaccine; Earth does not need to be thought as all anthropogenically constructed and consequently, dependent solely on human effort for survival and sustenance. We can cut down on plastic and yet cannot be free from this material as it builds its organic and synthetic ties with an erratic and transcendental Earth. The virus has its own ties with Earth as much as plastic has its own world of geo-negotiations to meet. A majority of us succumbs to plastic death, some mutate with plastic immunity, and some post-humanize under internal plastic deposition. Earth plasticizes its de-formations, lives out of a de-composition, and finds its exits in what we post-environmentalists would loosely qualify as anthropological post-endism. We struggle to keep pace with a "wandering" planet (*planasthai*, Greek, meaning "to wander")[12] and a plastic Earth beyond the human understanding of plastic—Earth under processes of discovery beyond a human discovery (plastic-material).

But how does that keep us alive? Amid the ever-mounting plastic cosmopolitan presence we aim to find a "sense of place": we live the inhuman time struggling with a disanthropic imagination.[13] Turning to plastic in production, use, and application—our inevitable co-passenger in everyday dynamics—is as inevitable as turning away from it. The end of plastic will be somewhere the end of Earth with greater carbon contamination and emission. Has the turn to plastic done only wrong to Earth and Earth life? Should Earth forget mankind?[14] Can a nonhuman Earth that is plastic-free, help the planet?

---

11. Neyrat, "Eccentric Earth," 6.
12. Neyrat, "Planetary Antigones," 36.
13. Garrard, "Worlds without Us."
14. See Gray, *Straw Dogs*. Thompson, "Plastics, Environment and Health," points out that the "use of plastic components in automobiles and aircraft results in significant weight savings compared with metals. The new Boeing 787 aircraft will, for example, have a skin that is 100 per

How would a post-human Earth look? Will it survive human extinction? The turn to plastic was a choice; turning away from plastic has left us with no choice. As a plastic species, we are all participants in a "common household of matter, desire, and imagination—an economy of metabolic and poetic transformations." Andreas Weber calls this "enlivenment."[15] This enlivenment is through plastic-connect—plastic inclusivity in our life-webbing—as the nature-culture divide collapses on the margins and axis of plastic live-liness. Plastic drive has brought us before plastic commons. "The term commons," Weber argues, "characterizes a form of socioeconomy that integrates material and emotional relationships. It is based on the exchange of goods, but also on transformations of meaning."[16] It now extends to ecology as plastic metabolism (en)folds all domains through its power of transformation and extensive invasive participation.

We are plastic-alive materially and aesthetically and material-aesthetically. Plastic thinking is living through the material and the aesthetic, the materiality and the imagination—a deep interaction with the matter and allowing a "dream" to process. The plastic turn makes for living poetically and dreaming materially. Gaston Bachelard notes that "the material imagination . . . goes beyond the attractions of the imagination of forms; [it] thinks matter, dreams in it, lives in it, or, in other words, materializes the imaginary."[17] The material-plastic element animates the mind—acting "like a good conductor that gives continuity to the imagining psyche"[18]—and sets off the imagination, the aesthetic, which is the "inner recesses of the matter" and

---

cent composite and an interior that is 50 per cent plastic, resulting in combined fuel savings of around 20 per cent. Single-use plastic packaging items are among the most common components of marine litter; however, such packaging has a key role in reducing wastage since even a relatively small use of packaging can extend the shelf life of perishable products, hence contributing to food safety. Since plastic packaging is lightweight, it can also achieve significant reductions in fuel usage (packaging in PET can achieve a 52 per cent saving over glass, for example) during transportation. It is this combined success that has led to global production of plastics, accounting for around 4 per cent of world oil production in the products themselves and a further 4 per cent in the energy required for this production. However, this success also results in the accumulation of end-of-life plastics that is being examined here. In order to generate solutions, an holistic framework is required, which aspires to optimize benefits and reduce impacts in order to harness the greatest potential that plastic products can offer humanity" (157).

15. Weber, *Enlivenment*, 2.
16. Weber, *Enlivenment*, 5.
17. Quoted in Harris, "Stoned Thinking," 140.
18. Bachelard, *Air and Dreams*, 8.

where "a material element must provide its own substance, its particular rules and poetics."[19] This, as Paul Harris leads me to see, is an "animaterialist imagination" that "exceeds the discourse of 'animating' that is ubiquitous in contemporary new materialisms."[20] This is a way of thinking matter—a "reverie" that is a deep listening to the work and the workability of the material. We are thrown into the plastic turn as a continuity and as an epiphanic instant, "listening" to the life of the inhuman, the subjective object, and the substance-ability of the material. Plastic speaks, has its own respiration, and is our modern-day reverie. The Plastic Turn, intriguingly, is with, without, and outside plastic.

---

19. Bachelard, *Water and Dreams*, 3.
20. Harris, "Stoned Thinking," 139.

BIBLIOGRAPHY

Adorno, Theodor. *Aesthetic Theory*. Translated by Robert Hullot-Kenner. Minneapolis: University of Minnesota Press, 1997.

Akrich, M., and Bruno Latour, "A Summary of a Convenient Vocabulary for the Semiotics of Human and Nonhuman Assemblies," In *Shaping Technology/Building Society*, edited by W. E. Bijker and J. Law, 259–269. Cambridge, MA: MIT Press, 1992.

Alaimo, Stacy. *Exposed: Environmental Politics and Pleasures in Posthuman Times*. Minneapolis: University of Minnesota Press, 2016.

Allured, Elizabeth. "Holding the Un-Grievable: A Psychoanalytic Approach to the Environmental Crisis. Review of *Climate Crisis, Psychoanalysis, and Radical Ethics*, by Donna M. Orange." *Contemporary Psychoanalysis* 54, no. 1 (2018): 239–246.

Amos, Jonathan. "Earth Is Becoming Planet Plastic." BBC, July 19, 2017. https://www.bbc.com/news/science-environment-40654915.

Anderson, Jon. "Relational Places: The Surfed Wave as Assemblage and Convergence." *Environment and Planning D: Society and Space* 30, no. 4 (2012): 576–587.

Anderson, Jon. "Transient Convergence and Relational Sensibility: Beyond the Modern Constitution of Nature." *Emotion, Space and Society* 2, no. 2 (2009): 120–127.

Andersson, I. K., and P. H. Kirkegaard. "A Discussion of the Term Digital Tectonics." In *Digital Architecture and Construction*, edited by A. Ali and C. A Brebbia (Southampton: WIT Press, 2006).

Andrady, Anthony L. "Microplastics in the Marine Environment." *Marine Pollution Bulletin* 62, no. 8 (2011): 1596–1605.

Anjos, Moacir dos. "Where All Places Are." In *Cildo Meireles*, edited by Guy Brett, 170–73. London: Tate Modern, 2008.

Anstead, David Thomas. *The Great Stone Book of Nature*. Philadelphia: George W. Childs, 1863.

Anthony, Carl. "Ecopsychology and the Deconstruction of Whiteness." In *Ecopsychology*, edited by T. Roszak, M. E. Gomes, and A. D. Kanner, 263–278. San Francisco: Sierra Club Books, 1995.

Araujo, Denize Correa. "Postmodernist Intertextuality: Artistic Clonage, Anthropophagic Pastiche, or What?" PhD diss., University of California, 1998.

Ardenne, Paul. "Ecological Art—Origins, Reality, Becoming." In *Art, Theory and Practice in the Anthropocene*, edited by Julie Reiss, 51–64. Wilmington, DE: Vernon Press, 2019.

Ashford, Molika. "What Happens Inside a Landfill?" *Live Science*, August 25, 2010. https://www.livescience.com/32786-what-happens-inside-a-landfill.html.

Auta, H. S., C. U. Emenike, and S. H. Fauziah. "Distribution and Importance of Microplastics in the Marine Environment: A Review of the Sources, Fate, Effects, and Potential Solutions." *Environment International* 102 (May 2017): 165–176.

Avery-Gomm, Stephanie, Stephanie B. Borrelle, and Jennifer F. Provencher. "Linking Plastic Ingestion Research with Marine Wildlife Conservation." *Science of the Total Environment* 637–638 (October 2018): 1492–1495.

Bachelard, Gaston. *Water and Dreams: An Essay on the Imagination of Matter*. Translated by Edith R. Farrell. Dallas, TX: Dallas Institute of Humanities and Culture Publications, 1999.

Bachelard, Gaston. *Earth and Reveries of Will: An Essay on the Imagination of Matter*. Translated by Kenneth Haltman. Dallas, TX: Dallas Institute of Humanities and Culture Publications, 2002.

Bachelard, Gaston. *Air and Dreams: An Essay on the Imagination of Movement*. Translated by Edith Farell and Frederick Farell. Dallas, TX: Dallas Institute of Humanities and Culture Publications, 2011.

Bacskay, George B., Jeffrey R. Reimers, and Sture Nordholm. "The Mechanism of Covalent Bonding." *Journal of Chemical Education* 74, no. 12 (1997): 1494.

Badenhausen, Richard. *T. S. Eliot and the Art of Collaboration*. Cambridge, UK: Cambridge University Press, 2004.

Badiou, Alan. *Being and Event*. Translated by Oliver Feltham. New York: Continuum, 2005.

Bahun, Sanja. "Politics of World Literature." In *Routledge Companion to World Literature*, edited by Theo D'haen, David Damrosch, and Djelal Kadir, 373–382. London: Routledge, 2012.

Bakhtin, Mikhail. *The Dialogic Imagination: Four Essays*. Edited by Michael Holquist. Translated by Caryl Emerson and Michael Holquist. Austin: University of Texas Press, 1981.

Ballard, Edward G. "The Nature of the Object as Experienced." *Research in Phenomenology* 6 (1976): 105–138.

Barbiero, Daniel. "The Object as Catalyst: The Enigmatic Mask." *Arteidolia*, May 2018. http://www.arteidolia.com/the-object-as-catalyst-the-enigmatic-mask/.

Barnes, David K. A., et al. "Accumulation and Fragmentation of Plastic Debris in Global Environments." *Philosophical Transactions of the Royal Society B: Biological Sciences* 364, no. 1526 (July 27, 2009): 1985–1998.

Barthes, Roland. "Theory of the Text." In *Untying the Text: A Post-Structuralist Reader,* edited by Robert Young, 31–47. London: Routledge and Kegan Paul, 1981.

Barthes, Roland. "Plastic." *Perspecta* 24 (1988): 92–93.

Bartky, Sandra Lee. "Heidegger and the Modes of World-Disclosure." *Philosophy and Phenomenological Research* 40, no. 2 (1979): 212–236.

Baudrillard, Jean. *Systems of Objects.* Translated by James Benedict. London: Verso, 2006.

Bauman, Richard. *A World of Other's World: Cross-Cultural Perspectives on Intertextuality.* Oxford: Blackwell, 2004.

Baumgarten, Murray. "Primo Levi's 'Small Differences' and the Art of *The Periodic Table*: A Reading of 'Potassium.'" *Shofar* 32, no. 1 (2013): 60–78.

Bell, Elizabeth S. "The Language of Discovery: The Seascapes of Rachel Carson and Jacques Cousteau." *CEA Critic* 63, no. 1 (2000): 5–13.

Bell, Matthew. "On Granite." In *The Essential Goethe.* Princeton, NJ: Princeton University Press, 2016.

Bennett, Jane. "Earthling, Now and Forever?" In *Making the Geologic Now: Responses to Material Conditions of Contemporary Life,* edited by Elizabeth Ellsworth and Jamie Kruse, 243–246. New York: Punctum Books, 2013.

Bensaude Vincent, Bernadette. "Reconfiguring Nature through Syntheses: From Plastics to Biomimetics." In *The Natural and the Artificial: An Ever-Evolving Polarity,* edited by B. Bensaude Vincent and W. R. Newman, 293–312. Cambridge, MA: MIT Press, 2007.

Bensaude Vincent, Bernadette. "Plastics, Materials and Dreams of Dematerialization." In *Accumulation: The Material Politics of Plastic,* edited by Jennifer Gabrys, Gay Hawkins, and Mike Michael, 17–29. London: Routledge, 2013.

Bensaude Vincent, Bernadette, and Isabelle Stengers. *A History of Chemistry.* Cambridge, MA: Harvard University Press, 1996

Berg, Stephen. "An Introduction to Oswald de Andrade's *Cannibalist Manifesto,*" *Third Text* 13, no. 46 (1999), 89–91.

Berleant, Arnold. "Ideas for a Social Aesthetic." In *The Aesthetics of Everyday Life,* edited by Andrew W. Light and Jonathan M. Smith, 26–29. New York: Columbia University Press, 2005.

Best, Susan. "Rethinking Visual Pleasure: Aesthetics and Affect." *Theory Psychology* 17, no. 4 (2007): 505–514.

Bhabha, Homi. "Adagio." *Critical Inquiry* 31, no. 2 (2005): 371–380.

Bigda-Peyton, Frances. "When Drives Are Dangerous: Drive Theory and Resource Overconsumption." *Modern Psychoanalysis* 29, no. 2 (2004): 251–270.

Birley, A. W., R. J. Heath, and M. J. Scott. *Plastic Materials: Properties and Applications.* Dordrecht, NL: Springer Netherlands, 1991.

Black, Max. "More about Metaphor." *Dialectica* 31, nos. 3–4 (1977): 431–457.

Blaszczyk, Regina Lee. "Chromophilia: The Design World's Passion for Colour." *Journal of Design History* 27, no. 3 (2014): 203–217.

Bloomfield, Mandy. "Widening Gyre: A Poetics of Ocean Plastics." *Configurations* 27, no. 4 (2019): 501–523.

Blum, Hester. "'Bitter with the Salt of Continents': Rachel Carson and Oceanic Returns." *WSQ: Women's Studies Quarterly* 45, nos. 1–2 (2017): 287–291.

Bocqué, Maëva, Coline Voirin, Vincent Lapinte, Sylvain Caillol, and Jean-Jacques Robin, "Petro-Based and Bio-Based Plasticizers: Chemical Structures to Plasticizing Properties." *Polymer Chemistry* 54, no. 11 (January 2016): 11–33.

Boetzkes, Amanda. *The Ethics of Earth Art.* Minneapolis: University of Minnesota Press, 2010.

Bök, Christian. "Virtually Nontoxic." *English Studies in Canada* 42, nos. 3–4 (2016): 25–26.

Bonta, M., and J. Protevi. *Deleuze and Geophilosophy: A Guide and Glossary.* Edinburgh: Edinburgh University Press, 2004.

Boomen, Marianne van den. "Interfacing by Iconic Metaphors." *Configurations* 16, no. 1 (2008): 33–55.

Borges, Jorge Louis. *Labyrinths: Selected Stories and Other Writing*, edited by D. A. Yates and J. E. Irby. New York: New Direction, 1964.

Bourriaud, Nicolas. *Altermodern.* London: Tate, 2009.

Braidotti, Rosi, and Rick Dolphijn, eds. *This Deleuzian Century: Art, Activism, Life.* Leiden: Brill Rodopi, 2014.

Breton, Andre. *L'amour fou.* Paris: Gallimard, 1937.

Brickey, Russell. "Poe's Dark Sublime and the Supernatural." *English Studies* 100, no. 7 (2019): 785–804.

Brown, Nathan. *Limits of Fabrication: Materials Science, Materialist Poetics.* New York: Fordham University Press, 2017.

Brugidou, Jeremie, and Clouette Fabien. "'AnthropOcean': Oceanic Perspectives and Cephalopodic Imaginaries Moving beyond Land-Centric Ecologies." *Social Science Information* 57, no. 3 (2018): 359–385.

Bryant, John. "Poe's Ape of UnReason: Humor, Ritual, and Culture." *Nineteenth Century Literature* 51, no. 1 (1996): 16–52.

Buchanan, Richard. "Anxiety, Wonder and Astonishment: The Communion of Art and Design." *Design Issues* 23, no. 4 (2007): 39–45.

Buck, Brennan. "What Plastic Wants." *Log*, no. 23 (2011): 35–40.

Buckland, Adelene. "Inhabitants of the Same World: The Colonial History of Geological Time." *Philological Quarterly* 97, no. 2 (2018): 219–240.

Caffentzis, George. *No Blood for Oil! Energy, Class Struggle, and War, 1998–2004.* New York: Autonomedia, 2017.

Caillois, Roger. *Pierres réfléchies.* Paris: Editions Gallimard, 1975.

Cameron, Dan, and Cildo Meireles. "Cildo Meireles." *BOMB* no. 70 (2000): 124–127.

Cantor, Kirk M., and Patrick Watts. "Plastics Processing." In *Applied Plastics Engineering Handbook*, edited by Myer Kutz, 195–203. Waltham, MA: Elsevier, 2011.

Capozzi, Rocco. "Palimpsests and Laughter: The Dialogical Pleasure of Unlimited Intertextuality in *The Name of the Rose*." *Italica* 66, no. 4 (1989): 412–428.

Capozzi, Rocco. "Intertextuality and the Proliferation of Signs/Knowledge in Eco's Super-Fictions." *Forum Italicum: A Journal of Italian Studies* 32, no. 2 (1998): 456–481.

Cardinal, Roger. "André Breton: The Surrealist Sensibility." *Mosaic: A Journal for the Interdisciplinary Study of Literature* 1, no. 2. (1968), 112–126.

Carroll, Noël. "On Being Moved by Nature." In *Arguing about Art: Contemporary Philosophical Debates*, 2nd ed., edited by Alex Neill and Aaron Ridley. London: Routledge, 2002.

Carroll, William F., Jr., Richard W. Johnson, Sylvia S. Moore, and Robert A. Paradis. "Poly(Vinyl Chloride)." In *Applied Plastics Engineering Handbook*, edited Myer Kutz, 73–90. Waltham, MA: Elsevier, 2011.

Carson, Rachel. *The Sea around Us*. Oxford: Oxford University Press, 2018.

Castro, Joseph. "Plastic Legacy: Humankind's Trash Is Now a New Rock." *Live Science*, June 3, 2014. https://www.livescience.com/46057-human-trash-becomes-new -plastiglomerate-rock.html.

Celina, Jeffery. "Artistic Immersion: Towards an Oceanic Connectedness." *Symploke* 27, no. 1–2 (2019): 35–46.

Chakrabarty, Dipesh. "The Climate of History: Four Theses." *Critical Inquiry* 35, no. 2 (2009): 197–222.

Chandler, Mielle, and Astrida Neimanis. "Water and Gestationality: What Flows beneath Ethics." In *Thinking with Water*, edited by Cecilia Chen, Janine MacLeod, and Astrida Neimanis, 61–83. Montreal: McGill-Queen's University Press, 2013.

Chang, Weiyi. "After Nature and Culture: Plastiglomerate in the Age of Capital." In *Art, Theory and Practice in the Anthropocene*, edited by Julie Reiss, 99–110. Delaware: Vernon Press, 2019.

Cheah, Pheng. "What Is a World?: On World Literature as World-Making Activity." *Daedalus* 137, no. 3 (2008): 26–38.

Chen, Cecilia, Janine MacLeod, and Astrida Neimanis. "Introduction: Toward a Hydrological Turn?" In *Thinking with Water*, edited by Cecilia Chen, Janine MacLeod, and Astrida Neimanis, 3–22. Montreal: McGill-Queen's University Press, 2013.

Chen, Guo-Neng, and Rodney Grapes. *Granite Genesis: In Situ Melting and Crustal Evolution*. Dordrecht, NL: Springer, 2007.

Chen, Mel Y. "Toxic Animacies, Inanimate Affections." *GLQ* 17, nos. 2–3 (2011): 265–286.

Chen Jun, Angus. "Rocks Made of Plastic Found on Hawaiian Beach." *Science Magazine*, June 4, 2014. https://www.sciencemag.org/news/2014/06/rocks-made-plastic -found-hawaiian-beach.

"Cildo Meireles, Babel, 2001." Tate, webpage. Accessed November 22, 2020. https://www .tate.org.uk/art/artworks/meireles-babel-t14041.

"Cildo Meireles." Art and Artworks webpage. Accessed November 22, 2020. https://www .artartworks.com/exhibitions/cildo-meireles-755/.

Clark, Samantha. "Contemporary Art and Environmental Aesthetics." *Environmental Values* 19, no. 3 (2010): 351–371.

Clarke, S., and P Hoggett, eds. *Object Relations and Social Relations: The Implications of the Relational Turn in Psychoanalysis*. London: Karnac, 2008.

Clement, C. *Syncope: The Philosophy of Rapture*. Minneapolis: University of Minnesota Press, 2004.

Colebrook, Claire. "Queer Vitalism." *New Formations* 68, no. 1 (2009): 77–92.

Colquhoun, Ithell. *The Living Stones: Cornwall*. London: Peter Owen, 2017.

Combes, Muriel. *Gilbert Simondon and the Philosophy of the Transindividual.* Translated by Thomas LaMarre. Cambridge, MA: MIT Press, 2013.

Compagnon, Antoine. *Les cinq paradoxes de la modernité.* Paris: Seuil, 1990.

Conley, Katharine. *Surrealist Ghostliness.* London: University of Nebraska Press, 2013.

Conley, Katharine. "Collecting Ghostly Things: André Breton and Joseph Cornell." *Modernism/Modernity* 24, no. 2 (2017): 263–282.

Coppan, Vilashini. "Codes for World Literature: Network Theory and the Field Imaginary." In *Approaches to World Literature*, edited by Joachim Küpper, 103–120. Berlin: Akademie Verlag, 2013.

Corcoran, Patricia L., Charles J. Moore, and Kelly Jazvac. "An Anthropogenic Marker Horizon in the Future Rock Record." *GSA TODAY* 24, no. 6 (June 2014): 4–8.

Cózar, Andrés, Fidel Echevarría, J. Ignacio González-Gordillo, Xabier Irigoien, Bárbara Úbeda, Santiago Hernández-León, Álvaro T. Palma, Sandra Navarro, Juan García-de-Lomas, Andrea Ruiz, María L. Fernández-de-Puelles, and Carlos M. Duarte. "Plastic Debris in the Open Ocean." *PNAS* 111, no. 28 (2014): 10239–10244. https://doi.org/10.1073/pnas.1314705111.

Cran, Rona. *Collage in Twentieth-Century Art, Literature, and Culture: Joseph Cornell, William Burroughs, Frank O'Hara, and Bob Dylan.* Farnham, Surrey: Ashgate, 2014.

Croissant, Jennifer L. "Agnotology: Ignorance and Absence or Towards a Sociology of Things That Aren't There." *Social Epistemology* 28, no. 1 (2014): 4–25.

Crowther T. "The Matter of Events." *Review of Metaphysics* 65, no. 1 (2011): 3–39.

Csernica, Jeffrey, and Alisha Brown. "Effect of Plasticizers on the Properties of Polystyrene Films." *Journal of Chemical Education* 76, no. 11 (Nov 1999): 1526.

Dahlsen, John. "Advice from the Ocean: Unexpected Paths into Marine Conservation." Accessed September 27, 2021. https://johndahlsen.com/latest-artist-statements/advice-from-the-ocean-unexpected-paths-into-marine-conservation/.

Dahlsen, John. "Environmental Art: Activism, Aesthetics and Transformation." PhD. diss., Charles Darwin University, 2017.

Dahlsen, John. "John Dahlsen Artist Statement 2019." Accessed November 23, 2020. https://johndahlsen.com/latest-artist-statements/john-dahlsen-artist-statement-2019/.

Dahlsen, John. "Marine Conservation: Going beyond Biology and Expanding into Art." Accessed November 23, 2020. https://johndahlsen.com/latest-artist-statements/advice-from-the-ocean-unexpected-paths-into-marine-conservation/.

Dahlsen, John. "Painting with an Environmental Palette." Accessed November 23, 2020. https://johndahlsen.com/interviews/painting-with-an-environmental-palette/.

Darwin, Charles. *Journal of Researches into the Natural History and Geology of the Countries Visited during the Voyage of H.M.S. "Beagle" around the World.* Cambridge: Cambridge University Press, 2011.

Davis, Heather. "Life & Death in the Anthropocene: A Short History of Plastic." In *Art in the Anthropocene: Encounters among Aesthetics, Politics, Environments and Epistemologies*, edited by Heather Davis and Etienne Turpin, 347–358. London: Open Humanities Press, 2015.

Davis, Heather. "Toxic Progeny: The Plastisphere and Other Queer Futures." *philoSOPHIA* 5, no. 2 (2015): 231–250.

de Andrade, Oswald. "The Cannibalist Manifesto." Translated by Leslie Bary. *Latin American Literary Review* 19, no. 38 (1991).

Degersdorff, Anne F. "Schizoid Phenomena, Object Relations and the Self." *Contemporary Psychoanalysis* 6, no. 1 (1969): 84–88.

Delaney, Rachael. "Other Lessons Learned from Relational Aesthetics." *Visual Arts Research* 40, no. 1 (2014): 25–27.

Deleuze, Gilles. *Cinema 2: The Time-Image*. Translated by Hugh Tomlinson and Robert Galeta. Minneapolis: University of Minnesota Press, 1989.

Deleuze, Gilles, and Felix Guattari, *A Thousand Plateaus: Capitalism and Schizophrenia*. Translated by Brian Massumi. Minneapolis: University of Minnesota Press, 2005.

DeLoughrey, E. "Towards a Critical Ocean Studies for the Anthropocene." *English Language Notes* 57, no. 1 (2019): 21–36.

DeLoughrey, Elizabeth. "Gyre." *Society & Space*, April 9, 2019. https://www.societyandspace.org/articles/gyre.

De Loughry, Treasa. "Polymeric Chains and Petrolic Imaginaries: World Literature, Plastic, and Negative Value." *Green Letters* 23, no. 2 (2019): 179–193.

Demos, Virginia, ed. *Exploring Affect: The Selected Writings of Silvan S. Tomkins*. New York: Cambridge University Press, 1995.

Devisch, Ignaas. "The Sense of Being(-)with Jean-Luc Nancy." *Culture Machine* 8 (2006). https://culturemachine.net/community/the-sense-of-being/.

Dewdney, Christopher. *The Secular Grail: Paradigms of Perception*. Toronto: Somerville House Books, 1993.

De Wolff, Kim. "Plastic Naturecultures: Multispecies Ethnography and the Dangers of Separating Living from Nonliving Bodies." *Body and Society* 23, no. 3 (2017): 23–47.

Dezeuze, Anna. "Cildo Meireles." *Artforum International* 47, no. 8 (2009): 182.

Di Chiro, Giovanna. "Polluted Politics? Confronting Toxic Discourse, Sex Panic, and Eco-Normativity." In *Queer Ecologies Sex, Nature, Politics, Desire*, edited by Catriona Mortimer-Sandilands and Bruce Erickson, 199–230. Indianapolis: Indiana University Press, 2010.

Dodds, Joseph. "Psychoanalysis and Ecology at the Edge of Chaos." MPhil thesis, Sheffield University, 2010.

Dodds, Joseph. *Psychoanalysis and Ecology at the Edge of Chaos: Complexity Theory, Deleuze, Guattari and Pscyhoanalysis for a Climate in Crisis*. New York: Routledge, 2011.

Dodds, Joseph. "Minding the Ecological Body: Neuropsychoanalysis and Ecopsychoanalysis." *Frontiers in Psychology* 4, no. 125 (2013): 125. https://doi.org/10.3389/fpsyg.2013.00125.

Dolphijn, Rick. "The Revelation of a World That Was Always Already There: The Creative Act as an Occupation." In *This Deleuzian Century: Art, Activism, Life*, edited by Rosi Braidotti and Rick Dolphijn. Leiden: Brill Rodopi, 2014.

Dooren, Thom von. *Flight Ways*. New York: Columbia University Press, 2014.

Doss, Erika. "Affect." *American Art* 23, no. 1 (2009): 9–11.

Dufrenne, Mikel. *The Notion of the A Priori*. Translated by Edward S. Casey. Evanston, IL: Northwestern University Press, 2009.

Duncan, Ian. "On Charles Darwin and the *Voyage of the Beagle*," *BRANCH: Britain, Representation and Nineteenth-Century History*, ed. Dino Franco Felluga. Accessed

September 27, 2021. http://www.branchcollective.org/?ps_articles=ian-duncan-on
-charles-darwina-and-the-voyage-of-the-beagle-1831–36.

DunLany, Melissa. "The Aesthetics of Waste: Michel Tournier, Agnes Varda, Sabina Macher." PhD diss., University of Pennsylvania, 2017.

Dunn, Christopher. *Brutality Garden: Tropicália and the Emergence of a Brazilian Counterculture.* Chapel Hill: University of North Carolina Press, 2001.

Dunn, Christopher. *Contracultura.* Chapel Hill: University of North Carolina Press, 2016.

Duran, Jessica Pujol. "Umberto Eco's New Paradigm and Experimentalism in the 1960s." *Zagadnienia Rodzajów Literackich* 58, no. 2 (2015): 51–61.

Dynisco. "Polymers and Plastics." AZoM, February 14, 2017. https://www.azom.com /article.aspx?ArticleID=13567.

Earley, Joseph E. "Some Philosophical Implications of Chemical Symmetry." In *Philosophy of Chemistry: Synthesis of a New Discipline*, edited by Davis Baird, Eric Scerri, and Lee McIntyre, 207–220. Dordrecht, NL: Springer, 2006.

Eco, Umberto. *Semiotics and Philosophy of Language.* Bloomington: Indiana University Press, 1984.

Eco, Umberto. *The Aesthetics of Chaosmos. The Middle Ages of James Joyce.* Translated by Ellen Esrock. Cambridge, MA: Harvard University Press, 1989.

Efal, Adi. "Gravity of a Figure." *Journal of Visual Art Practice* 12, no. 1 (2013): 39–49.

"Elastomers." Polymer Properties Database. Accessed June 21, 2021. https://polymerdatabase .com/Elastomers/Elastomers.html.

Elias, Amy J. "The Dialogical Avant-Garde: Relational Aesthetics and Time Ecologies in Only Revolutions and TOC." *Contemporary Literature* 53, no. 4 (2012): 738–778.

Elias, Amy J., and Christian Moraru, eds. *The Planetary Turn: Relationality and Geoaesthetics in the Twenty-First Century.* Evanston, IL: Northwestern University Press, 2015.

Eliot, T. S. *The Waste Land.* Edited by Michael North. New York: W. W. Norton, 2001.

Epps, Philomena. "Performance at Tate: Into the Space of Art: Hélio Oiticica, Parangolés 2007." Tate, April 2016. https://www.tate.org.uk/research/publications/performance-at -tate/case-studies/helio-oiticica.

Eriksen, Marcus. "The Plastisphere: The Making of a Plasticized World." *Tulane Environmental Law Journal* 27, no. 1 (2014): 153–162.

Eriksen, Marcus. *Junk Raft.* Boston: Beacon Press, 2018.

Evenson, Edward B., Patrick A. Burkhart, John C. Gosse, Gregory S. Baker, Dan Jackofsky, Andres Meglioli, Ian Dalziel, Stefan Kraus, Richard B. Alley, and Claudio Berti. "Enigmatic Boulder Trains, Supraglacial Rock Avalanches, and the Origin of 'Darwin's Boulders,' Tierra del Fuego." *GSA TODAY* 19, no. 12 (2009). https://www .geosociety.org/gsatoday/archive/19/12/article/i1052-5173-19-12-4.htm.

"Evidence That the Great Pacific Garbage Patch Is Rapidly Accumulating Plastic." *Scientific Reports* 8, no. 4666 (2018), https://doi.org/10.1038/s41598-018-22939-w.

Farmer, John Alan. "Through the Labyrinth: An Interview with Cildo Meireles." *Art Journal* 59, no. 3 (2000): 34–43.

Fenner, David E. W. "Aesthetic Experience and Aesthetic Analysis." *Journal of Aesthetic Education* 37, no. 1 (2003): 40–53.

Fisher, A. "Ecopsychology at the Crossroads: Contesting the Nature of a Field." *Ecopsychology* 5, no. 3 (2013): 167–176.

Fisher, Andy. "Going Deep: A Review of *Environmental Melancholia: Psychoanalytic Dimensions of Engagement* by Renee Lertzman." *Ecopsychology* 8, no. 4 (2016): 222–227.

Fisher, Tom H. "What We Touch, Touches Us: Materials, Affects, and Affordances." *Design Issues* 20, no. 4 (2004): 20–31.

Focillon, Henri. *The Life of Forms in Art.* Translated by C. B. Hogan and G. Kubier. New York: Zone Books, 1992.

Foster, Dennis A. "Re-Poe Man: Poe's Un-American Sublime." In *Sublime Enjoyment: On the Perverse Motive in American Literature*, 38–66. New York: Cambridge University Press, 1997.

Fraga, Mari. "Fossil Time: Oil, Art and Body in the Cosmopolitics of the Anthropocene." *Brazilian Journal of Presence Studies* 8, no. 1 (2018): 31–62. doi:10.1590 / 2237-266072285.

Frampton, Kenneth. "Rappel á l'ordre, the Case for the Tectonic (1995)." In *Theorizing a New Agenda for Architecture: An Anthology of Architectural Theory 1965–1995*, edited by Kate Nesbitt, 518–528. New York: Princeton Architectural Press, 1996.

Friedel, R. *Pioneer Plastic: The Making and Setting of Celluloid.* Madison: University of Wisconsin Press, 1983.

Friedman, Susan. *Planetary Modernism: Provocations of Modernity across Time.* New York: Columbia University Press, 2015.

Gablik, Suzi, "Connective Aesthetics." *American Art* 6, no. 2 (1992): 2–7.

Gablik, Suzi. "Connective Aesthetics: Art after Individualism." In *Mapping the Terrain: New Genre Public Art*, edited by Suzanne Lacy, 74–87. Seattle: Bay Press, 1995.

Galison, Peter. "Einstein's Clocks: The Place of Time." *Critical Inquiry* 26, no. 2 (2000): 355–389.

Galloway, Alexander R. "Plastic Reading." *Forum on Fiction* 45, no. 1 (2012): 10–12.

Galton, A. "The Ontology of States, Processes, and Events." *Interdisciplinary Ontology* 5, no. 1 (2012): 35–45.

Garcia, Cynthia. "The Scale of Common Things: Conceptual Giant Cildo Meireles Mounts Impressive Survey." *New City Brazil*, November 5, 2019. https://www .newcitybrazil.com/2019/11/05/the-scale-of-common-things-conceptual-giant-cildo -meireles-mounts-impressive-survey/.

Garrard, Greg. "Worlds without Us: Some Types of Disanthropy." *SubStance* 41, no. 127 (2012): 40–60.

Gee, Gabriel N. "Nature, Plastic, Artifice in Conversation with Tuula Närhinen." In *Changing Representations of Nature and the City: The 1960s–1970s and Their Legacies*, edited by Gabriel N. Gee and Alison Vogelaar,112–117. New York: Routledge, 2019.

Genette, Gerard. *Palimpsests: Literature in the Second Degree.* Translated by Channa Newman and Claude Doubinsky. Lincoln: University of Nebraska Press, 1997.

Gerhardt, Christina. "Pacific and Plastic: Midway Atoll, Plastiglomerate, and Love of Place." *Mosaic: An Interdisciplinary Critical Journal* 51, no. 3 (2018): 123–140.

Gersdorff, Anne F. de. "Schizoid Phenomena, Object Relations and the Self." *Contemporary Psychoanalysis* 6, no. 1 (1969): 84–88.

Ghosh, Ranjan. "Globing the Earth: The New Eco-Logics of Nature." *SubStance* 41, no. 1 (2012): 3–14.

Ghosh, Ranjan. "Aesthetic Imaginary: Rethinking the Comparative." *Canadian Review of Comparative Literature* 44, no. 3 (2017): 449–467.

Ghosh, Ranjan. *Transcultural Poetics and the Concept of the Poet.* New York: Routledge, 2017.

Ghosh, Ranjan. "Plastic Literature." *University of Toronto Quarterly* 88, no. 2 (2019): 277–291.

Ghosh, Ranjan. *Trans(in)fusion: Reflections for Critical Thinking.* New York: Routledge, 2020.

Ghosh, Ranjan, and J. Hillis Miller. *Thinking Literature across Continents.* Durham, NC: Duke University Press, 2016.

Ghosh, Ranjan, and Ethan Kleinberg, eds. *Presence: Philosophy, History, and Cultural Theory for the Twenty-First Century.* Ithaca, NY: Cornell University Press, 2013.

Giunta, Andrea. "Brazilian Art under Dictatorship: *Antonio Manuel, Artur Barrio, and Cildo Meireles* by Claudia Calirman (review)." *The Americas* 69, no. 4 (2013): 533–536.

Gloag, John. "The Influence of Plastics on Design." *Journal of the Royal Society of Arts* 91, no. 4644 (1943): 461–470.

Godwin, Allen D. "Plasticizers." In *Applied Plastics Engineering Handbook,* edited by Myer Kutz, 533–553. Waltham, MA: Elsevier, 2011.

Goethe, Johann Wolfgang von. "On Granite." In *The Essential Goethe,* edited by Matthew Bell, 2061–2067. Princeton, NJ: Princeton University Press, 2016.

Grant, Ian Hamilton. *Philosophies of Nature after Shelling.* London: Bloomsbury, 2006.

Gray, John. *Stray Dogs: Thoughts on Humans and Other Animals.* London: Granta, 2003.

Gregory, Joshua C. "Cudworth and Descartes." *Philosophy* 8, no. 32 (1933): 454–467.

Gregory, Murray R. "Environmental Implications of Plastic Debris in Marine Settings—Entanglement, Ingestion, Smothering, Hangers-On, Hitch-Hiking and Alien Invasions." *Philosophical Transactions of the Royal Society B: Biological Sciences* 364, no. 1526 (July 27, 2009): 2013–2025.

Greimas, Algirdas Julien, Frank Collins, and Paul Perron. "Figurative Semiotics and the Semiotics of the Plastic Arts." *New Literary History* 20, no. 3 (1989): 627–649.

Griffiths, Matthew. "Jorie Graham's Sea Change: The Poetics of Sustainability and the Politics of What We're Sustaining." In *Literature and Sustainability,* edited by Adeline Johns-Putra, John Parham, and Louise Squire, 221–228. Manchester: Manchester University Press, 2017.

Grosz, Elizabeth. "Habit Today: Ravaisson, Bergson, Deleuze and Us." *Body and Society* 19, nos. 2–3 (2013): 217–239.

Groves, Jason. "Goethe's Petrofiction: Reading the *Wanderjahre* in the Anthropocene." *Goethe Yearbook* 22 (2015): 95–113.

Guntrip, Harry. *Schizoid Phenomena, Object Relations, and the Self.* New York: International Universities Press, 1969.

Hamilton, Clive. *Defiant Earth: The Fate of Humans in the Anthropocene.* Crows Nest, NSW: Allen & Unwin, 2017.

Haraway, Donna J. *Staying with the Trouble: Making Kin in the Chthulucene.* Durham, NC: Duke University Press, 2016.

Haraway, Donna J., and Thyrza Nichols Goodeve. *How Like a Leaf: An Interview with Thyrza Nichols Goodeve.* New York: Routledge, 2000.

Hardin, Garrett. "From 'The Tragedy of the Commons.'" In *The Oxford Book of Modern Science Writing,* edited by Richard Dawkins, 263–266. New York: Oxford University Press, 2008.

Hardt, M. "Foreword: What Affects Are Good For." In *The Affective Turn: Theorizing the Social*, edited by P. T. Clough and J. Halley, ix–xiii. Durham, NC: Duke University Press, 2007.

Harper, Charles A. *Handbook of Plastic Processes*. Hoboken, NJ: Wiley InterScience, 2006.

Harris, Paul A. "Stoned Thinking: The Petriverse of Pierre Jardin." *SubStance* 47, no. 2 (2018): 119–148.

Harrison, Robert. *Gardens: An Essay on the Human Condition*. Chicago: University of Chicago Press, 2008.

Harrison, Robert Pogue. *Juvenescence: A Cultural History of Our Age*. Chicago: University of Chicago Press, 2014.

Hayes, Nick. *The Rime of the Modern Mariner*. London: Jonathan Cape, 2011.

Hayles, N. Katherine. "The Transformation of Narrative and the Materiality of Hypertext." *Narrative* 9, no. 1 (2001): 21–39.

Haynes, W. *Men, Money and Molecules*. New York: Doubleday, Doran & Co., 1936.

Hazel. "Meshes of Freedom at Tate Modern." *Londonist*, October 22, 2008. https://londonist.com/2008/10/meshes_of_freedom_at_tate_modern.

Helmreich, Stefan. *Alien Ocean: Anthropological Voyages in Microbial Seas*. Berkeley: University of California Press, 2009.

Helmreich, Stefan. "Nature/Culture/Seawater." *American Anthropologist* 113, no. 1 (2011): 132–144.

Hempel, C. G. "Explanation in Science and in History." In *Frontiers of Science and Philosophy*, edited by R. G. Colodny, 9–33. Pittsburgh, PA: University of Pittsburgh Press, 1962.

Heringman, Noah. *Romantic Rocks, Aesthetic Ideology*. Ithaca, NY: Cornell University Press, 2010.

Herkenhoff, Paulo. *Cildo Meireles, Geografia do Brasil*. Rio de Janeiro: Artviva, 2001.

Higgitt, Catherine, Marika Spring, and David Saunders. "Pigment-Medium Interactions in Oil Paint Films Containing Red Lead or Lead-Tin Yellow." *National Gallery Technical Bulletin* 24 (2003): 75–95.

Hoffmann, Roald. "Molecular Beauty." *Journal of Aesthetics and Art Criticism* 48, no. 3 (1990): 191–204.

Hoffmann, Roald. "The Poetry of Molecules." In *We Are All Stardust: Scientists Who Shaped Our World Talk about Their Work, Their Lives and What They Still Want to Know*, edited by Stefan Klein, translated by Ross Benjamin, 139–164. New York: The Experiment, 2015.

Holy Upanishads—Brihadaranyaka Upanishad. Translated by Swami Nikhilananda. Accessed November 22, 2020. http://www.ishwar.com/hinduism/holy_upanishads/brihadaranyaka_upanishad/part_05.html.

Home, Stephen. "Cildo Meireles: The Gold Thread." *Third Text* 14, no. 52 (2000): 31–44.

Honnery, Rachel. "Absolute Kippleization and the Plastosystem: Metaphors to Address Complex Science in the Age of the Anthropocene." PhD diss., University of New South Wales, 2018.

Hood, Ernie. "Are EDCs Blurring Issues of Gender?" *Environmental Health Perspectives* 113, no. 10 (2005): A670–A677. https://www.ncbi.nlm.nih.gov/pmc/articles/PMC1281309/.

Hörl, Erich. *General Ecology: The New Ecological Paradigm*. London: Bloomsbury Academic, 2017.

Horn, Eva. "The Anthropocene Sublime: Justin Guariglia's Artwork." In *Art, Theory and Practice in the Anthropocene*, edited by Julie Reiss, 1–8. Delaware: Vernon Press, 2019.

Howard, Don. "Passion at a Distance." In *Quantum Reality, Relativistic Causality, and Closing the Epistemic Circle*, edited by W. C. Myrvold and J. Christian, 3–11. New York: Springer, 2009.

Huang, Michelle. "Ecologies of Entanglement in the Great Pacific Garbage Patch." *Journal of Asian American Studies* 20, no. 1 (February 2017): 95–117.

Hughes, R. "Riven: Badiou's Ethical Subject and the Event of Art as Trauma." *Postmodern Culture* 17, no. 3 (2007). doi:10.1353/pmc.2008.0005.

Hughes, Robert. "Bernard Stiegler, Philosophical Amateur, or, Individuation from Eros to Philia." *Diacritics* 42, no.1 (2014): 46–69.

Hussin, Nuriziani. "The Effects of Cross Linking Byproducts on the Electrical Properties of Low Density Polyethylene." PhD diss., University of Southampton, 2011.

Huyssen, Andreas. "Modernism at Large." In *Modernism*, edited by Astradur Eysteinsson and Vivian Liska, 53–66. Amsterdam: John Benjamins Publishing, 2007.

Ingold, Tim. "Toward an Ecology of Materials." *Annual Review of Anthropology* 41 (2012): 427–442.

Intemann, Nicole, and Julia Patschorke. *Plastian, der Kleine Fisch: . . . und wie er mit seinen Freunden auf einer abenteuerlichen Reise die Welt ein bisschen besser macht*. Munich: Oekom, 2017.

Itkonen, Esa. *Analogy as Structure and Process Approaches in Linguistics, Cognitive Psychology and Philosophy of Sciences*. Amsterdam: John Benjamins Publishing, 2005.

Ivar do Sul, Juliana A. and Monica F. Costa. "The Present and Future of Microplastic Pollution in the Marine Environment." *Environmental Pollution* 185 (2014): 352–364.

Iversen, Margaret. "Readymade, Found Object, Photograph." *Art Journal* 63, no. 2 (2004): 44–57.

Jackson, Mark. "Plastic Islands and Processual Grounds: Ethics, Ontology, and the Matter of Decay." *Cultural Geographies* 20, no. 2 (2013): 205–224.

Jagodzinski, Jan. "Into the Dark Blue: A Medi(t)ation on the Oceans—Its Pain, Its Wonder, Its Wild, and Its Hope." *Symploke* 27, nos. 1–2 (2019): 111–138.

Jain, Manju. *A Critical Reading of the Selected Poems of T. S. Eliot*. Delhi: Oxford University Press, 1991.

Jansen, Jeffrey A. "Plastics—It's All about Molecular Structure." *Plastics Engineering* 72, no. 8 (2016): 44–49.

Johnston, M. "Hylomorphism." *Journal of Philosophy* 103, no. 12 (2006): 652–698.

Jones, R. *Soft Machines*. Oxford: Oxford University Press, 2004.

Joyce, James. *Finnegan's Wake*. Oxford: Oxford University Press, 2012.

"Judith & Richard Lang." Oceanic Global. Accessed November 23, 2020. https://oceanic.global/judith-richard-lang/.

Juelskjaer, Malou, and Nete Schwennesen. "Intra-Active Entanglements: An Interview with Karen Barad." *Kvinder, Kon & Forskning*, nos. 1–2 (2012): 10–23.

Kainulainen, Maggie. "Saying Climate Change: Ethics of the Sublime and the Problem of Representation." *Symploke* 21, nos. 1–2 (2013): 109–123.

Kaup, Monika. *Mad Intertextuality: Madness in Twentieth-Century Women's Writing.* Trier: WVT, 1993.

Kassouf, Susan. "Psychoanalysis and Climate Change: Revisiting Searles's The Nonhuman Environment, Rediscovering Freud's Phylogenetic Fantasy, and Imagining a Future." *American Imago* 74, no. 2 (2017): 141–171.

Kearney, Richard. "Bachelard and the Epiphanic Instant." *Philosophy Today* 52 (2008): 38–45.

Kenyon, Karl W., and Eugene Kridler. "Laysan Albatrosses Swallow Indigestible Matter." *Auk* 86, no. 2 (1969): 339–343.

Kidner, David W. "Depression and the Natural World: Towards a Critical Ecology of Psychological Distress." *International Journal of Critical Psychology* 19 (2007): 123–146.

Kitts, David B. "Geologic Time." *Journal of Geology* 74, no. 2 (1966): 127–146.

Kłos, Anna. "MERZ—Kurt Schwitters' Combination of Art and Life." *Retroavangarda,* May 21, 2016. https://retroavangarda.com/merz-kurt-schwitters-2/.

Krauss, Rosalind. "In the Name of Picasso." *October* vol. 16, Art World Follies (Spring 1981), 5–22.

Kress, J., L. Clarkson, D. Moss, and L. Zeavin. "The Boston Declaration on the Role of the International Psychoanalytical Association in Addressing Global Change." *APSA,* July 25, 2015. http://www.apsa. org/content/news-international-psychoanalytical-associations-declaration-addressing-global-change.

Kronen, John, and Jacob Tulle. "Composite Substances as True Wholes: Toward a Modified Nyaya-Vaisesika Theory of Composite Substances." *Canadian Journal of Philosophy* 41, no. 2 (2011): 286–316.

Kristeva, Julia. *Desire in Language: A Semiotic Approach to Literature and Art.* Edited by Leon S. Roudiez. Translated by Thomas Gora, Alice Jardine, and Leon S. Roudiez. New York: Columbia University Press, 1980.

Lang, Richard, and Judith Selby Lang. *Love Hand.* In *Forever Plastic* blog, May 13, 2015. https://foreverplastic.wordpress.com/2015/05/13/love-hand/.

Lang, Richard, and Judith Selby Lang. The *Dinosaur* and the *Lamb.* In *Forever Plastic* blog, May 17, 2015. https://foreverplastic.wordpress.com/2015/05/17/dinosaur/.

Lang, Richard, and Judith Selby Lang. *Skeleton.* In *Forever Plastic* blog, May 17, 2015. https://foreverplastic.wordpress.com/2015/05/17/skeleton/.

Lang, Richard, and Judith Selby Lang. *The Great Wave.* In *Forever Plastic* blog, January 7, 2020. https://plasticforever.blogspot.com/2020/01/the-great-wave.html.

Lang, Richard, and Judith Selby Lang. *Plasticene Discontinuity.* In *Forever Plastic* blog. Accessed November 24, 2020. http://plasticforever.blogspot.com/p/exhibition-plasticene -discontinuity.html.

Lang, Richard, and Judith Selby Lang. "Interview." Oceanic Global. Accessed September 27, 2021. https://oceanic.global/judith-richard-lang/.

Langer, Ewa, Krzysztof Bortel, Marta Lenartowicz-Klik, and Sylwia Waskiewicz. *Plasticizers Derived from Post-Consumer Pet: Research Trends and Potential Applications.* Norwich, NY: William Andrew, 2019.

Lapworth, Andrew. "Habit, art, and the Plasticity of the Subject: The Ontogenetic Shock of the Bioart Encounter." *Cultural Geographies* 22, no. 1 (2015): 85–102.

Laszlo, Pierre. "Playing with Molecular Models." *Hyle—International Journal for Philosophy of Chemistry* 6 (2000): 85–97.

Latour, Bruno. *The Pasteurization of France.* Translated by Alan Sheridan and John Lee. Cambridge, MA: Harvard University Press, 1993.

Law, Kara Lavender, Skye Morét-Ferguson, Nikolai A. Maximenko, Giora Proskurowski, Emily E. Peacock, Jan Hafner, and Christopher M. Reddy. "Plastic Accumulation in the North Atlantic Subtropical Gyre." *Science* 329, no. 5996 (2010): 1185–1188.

Lehmann, Ann-Sophie. "How Materials Make Meaning." In *Nederlands Kunsthistorisch Jaarboek/ Netherlands Yearbook for History of Art* 62 (2012): 6–27.

Lehman, J. "A Sea of Potential: A Politics of Global Ocean Observations." *Political Geography* 55 (2016): 113–123.

Leigh, Nicola. *Albert Ross, The Albatross, Blue Spaghetti.* Self-published, Amazon Digital Services, 2018.

Lejeune, Denis. *The Radical Use of Chance in 20th Century Art.* Amsterdam: Rodopi, 2012.

Lemmens, Pieter. "This System Does Not Produce Pleasure Anymore: An Interview with Bernard Stiegler." *Krisis* no. 1 (2011): 33–41.

Lertzman, Renee Aron. "The Myth of Apathy: Psychoanalytic Explorations of Environmental Subjectivity." In *Engaging with Climate Change: Psychoanalytic and Interdisciplinary Perspectives,* edited by Sally Weintrobe, 117–133. New York: Routledge, 2013.

Lertzman, Renee Aron. *Environmental Melancholia: Psychoanalytic Dimensions of Engagement.* New York: Routledge, 2015.

Leservot, Typhaine. "From Weltliteratur to World Literature to Literature-Monde: The History of a Controversial Concept." In *Postcolonialism and Literature-monde,* edited by Alec G. Hargreaves, Charles Forsdick, and David Murphy, 36–48. Liverpool: Liverpool University Press, 2010.

Leslie, Esther. *Synthetic World: Nature, Art, and the Chemical Industry.* London: Reaktion, 2007.

Levi, Primo. *The Periodic Table.* Translated by Raymond Rosenthal. New York: Schocken, 1984.

Liboiron, Max. "Plastics in the Wild." PhD diss., New York University, 2012.

Liboiron, Max. "Plasticizers: A Twenty-First-Century Miasma." In *Accumulation: The Material Politics of Plastic,* edited by. J. Gabrys, G. Hawkins, and M. Michael, 140–141. London: Routledge, 2013.

Liehsi, Su. "The Bubbled Plastic Print: A New Approach to Printmaking." *Leonardo* 21, no. 4 (1988): 425–428.

Lockwood, Julie L. Martha F. Hoopes, and Michael P. Marchetti, *Invasion Ecology.* Oxford: Blackwell Publishing, 2007.

Lovett, Richard A. "Darwin's Geological Mystery Solved." *Scientific American,* October 20, 2009. https://www.scientificamerican.com/article/darwins-geological-mystery-sol/.

Lyell, Charles. *Principles of Geology.* Edited by James A. Secord. London: Penguin, 1997.

Malabou, Catherine. *The Future of Hegel: Plasticity, Temporality and Dialectic.* New York: Routledge, 2005.

Malabou, Catherine. "La generation d'apres." In *Fresh Theorie,* edited by Mark Alizart and Christophe Kihm, 539–553. Paris: Scheer, 2005.

Malabou, Catherine. *What Should We Do with Our Brain?* New York: Fordham University Press, 2008.

Malabou, Catherine. *Plasticity at the Dusk of Writing: Dialectic, Destruction, Deconstruction.* Translated by Carolyn Shread. New York: Columbia University Press, 2010.

Malabou, Catherine. *Ontology of the Accident: An Essay on Destructive Plasticity.* Malden, MA: Polity, 2012.

Malevich, Kazimir. "Spatial Cubism." In K. S. Malevich, *Essays on Art, 1915–1933*, vol. 2. Edited by Troels Andersen, 2–60. London: Rapp & Whiting, 1969.

Malkin, A. V., and G. V. Vinogradov. *Rheology of Polymers: Viscoelasticity and Flow of Polymers.* Berlin: Springer, 2013.

Marks-Tarlow, T. "The Self as a Dynamical System." *Nonlinear Dynamics, Psychology, and Life Sciences* 3, no. 4 (1999): 311–345.

Marshall, George. "The Psychology of Denial: Our Failure to Act Against Climate Change." *The Ecologist,* September 22, 2001. http://ecoglobe.ch/ motivation/e/clim2922.htm.

Martins, Sergio Bruno. "Hélio Oiticica: Mapping the Constructive." *Third Text* 24, no. 4 (2010): 409–442.

Martyn, Kevin P., and M. Martin Bosman. "Post-Truth or Agnogenesis? Theorizing Risk and Uncertainty in a Neoliberal Nature." *Journal of Risk Research* 22, no. 8 (2019): 951–963.

Marx, William. "The 20th Century: Century of the Arrière-Gardes?" In *Europa! Europa?: The Avant-Garde, Modernism and the Fate of a Continent,* edited by Sascha Bru, Jan Baetens, Benedikt Hjartarson, Peter Nicholls, Tania Ørum, and Hubert van den Berg. Berlin: Walter de Gruyter, 2009.

Masatoshi, Nakajima, and Tomii Reiko. "Readings in Japanese Art after 1945." In *Japanese Art after 1945: Scream against the Sky,* edited by Alexandra Munroe, 369–392. New York: Abrams, 1994.

McBride, Patrizia C., Richard W. McCormick, and Monika Žagar, eds., *Legacies of Modernism: Art and Politics in Northern Europe, 1890–1950.* N.p.: Palgrave Macmillan, 2007.

McKay, Micah D. "Stories from the Dump: Contemporary Latin American Trash Narratives." PhD diss., University of Wisconsin–Madison, 2017.

McKeen, Laurence. *The Effect of Sterilization Methods on Plastics and Elastomers.* Norwich, NY: William Andrew, 2018.

McKinnon, Andrew M. "Elective Affinities of the Protestant Ethic: Weber and the Chemistry of Capitalism." *Sociological Theory* 28, no. 1 (March 2010): 108–126.

McMenamin, Mark A. S., and Diana L. S. MacMenamin. *Hypersea: Life on Land.* New York: Columbia University Press, 1994.

Meikle, Jeffrey L. "Plastic, Material of a Thousand Uses." In *Imagining Tomorrow: History, Technology, and the American Future,* edited by J. Corn, 77–96. Cambridge, MA: MIT Press, 1986.

Meikle, Jeffrey L. "Into the Fourth Kingdom: Representations of Plastic Materials, 1920–1950." *Journal of Design History* 5, no. 3 (1992): 173–182.

Meikle, Jeffrey L. "Beyond Plastics: Postmodernity and the Culture of Synthesis." In *Ethics and Aesthetics: The Moral Turn of Postmodernism,* edited by Gerhard Hoffmann and Alfred Hornung, 325–342. Heidelberg: C. Winter, 1996.

Meikle, Jeffrey L. *American Plastic: A Cultural History.* New Brunswick, NJ: Rutgers University Press, 1997.

Meireles, Cildo. "Artist's Writings." In *Cildo Meireles.* London: Phaidon Press, 1999.

Meireles, Cildo. *Geografia do Brasil.* Rio de Janeiro: Artviva, 2001.

Meireles, Cildo, and Charles Merewether. "Memory of the Senses." *Grand Street* 64 (1998): 216–223.

Meltzer, Donald. *Psychoanalytical Process.* London: Harris Meltzer Trust, 2012.

Mertens, Heike Catherina. *Pinar Yoldas: An Ecosystem of Excess.* Berlin: Ernst Schering Foundation, 2014.

Metcalf, Robert. "The Elemental Sallis: On Wonder and Philosophy's 'Beginning.'" *Journal of Speculative Philosophy* 27, no. 2 (2013): 208–215.

Michael, Mike. "Futures of the Present: From Performativity to Prehension." In *Contested Futures,* edited by N. Brown, B. Rappert, and A. Webster, 21–39. Aldershot, UK: Ashgate, 2000.

Milman, Oliver. "'Great Pacific Garbage Patch' Far Bigger than Imagined, Aerial Survey Shows." *The Guardian,* October 4, 2016. https://www.theguardian.com /environment/2016/oct/04/great-pacific-garbage-patch-ocean-plastic-trash.

Mitchell, Audra. "Thinking without the Circle: Marine Plastic and the Global Ethics." *Political Geography* 47 (2015): 77–85.

Mitchell, Stephan A. *Relational Concepts within Psychoanalysis: An Integration.* Cambridge, MA: Harvard University Press, 1988.

Mitchell, Timothy. *Carbon Democracy: Political Power in the Age of Oil.* London: Verso, 2011.

Mitchell, William J. "Antitectonics: The Poetics of Virtuality." In *The Virtual Dimension: Architecture, Representation, and Crash Culture,* edited by John Beckman, 205–217. New York: Princeton Architectural Press, 1998.

Monbiot, George. "Climate Change: A Crisis of Collective Denial?" *Environmental Law and Management* 17, no. 2 (2005): 57–61.

Monbiot, George. *Relational Psychoanalysis.* N.p.: Analytic Press, 2005.

Moore, Charles, with Cassandra Phillips. *Plastic Ocean: How a Sea Captain's Chance Discovery Launched a Determined Quest to Save the Oceans.* Garden City, NY: Avery Penguin, 2011.

Moore, Jason. *Capitalism in the Web of Life: Ecology and the Accumulation of Capital.* London: Verso, 2015.

Moran, M. "Chaos and Psychoanalysis: The Fluidic Nature of Mind." *International Review of Psychoanalysis* 18, no. 2 (1991): 211–221.

Morton, Timothy. *Ecology without Nature.* Cambridge, MA: Harvard University Press, 2007.

Morton, Timothy. *Hyperobjects: Philosophy and Ecology after the End of the World.* Minneapolis: University of Minnesota Press, 2013.

Moser, Keith. "The Eco-Philosophy of Michel Serres and J. M. G. Le Clézio: Launching a Battle Cry to Save the Imperiled Earth." *Interdisciplinary Studies in Literature and Environment* 21, no. 2 (2014): 413–440.

Mossman, S., and Morris, P. eds. *The Development of Plastics.* London: The Science Museum, 1994.

Mulder, Jesse. "A Vital Challenge to Materialism." *Philosophy* 91, no. 2 (2016): 153–182.

Nancy, Jean-Luc. *Multiple Arts: The Muses II.* Stanford, CA: Stanford University Press, 2006.

Nancy, Jean-Luc, and Aurelien Barrau, *What's These Worlds Coming To?* New York: Fordham University Press, 2015.

Närhinen, Tuula. *Frutti di Mare.* 2008. http://www.tuulanarhinen.net/artworks/frutti .htm.

Närhinen, Tuula. "Between the City and the Deep Sea: On the Plastic Nature of the Helsinki Shoreline." In *Maritime Poetics: From Coast to Hinterland*, ed. Gabriel N. Gee and Caroline Wiedmer. Bielefeld, Germany: transcript Verlag, 2021. ttps://doi.org/10 .14361/9783839450239.

Närhinen, Tuula. "True Colours of Twilight—Twisting Strands of Experiment and Experience in the Fabric of Art and Science." In *I Experience as I Experiment—I Experiment as I Experience: Experience and Experimentality in Artistic work and Research*, edited by Denise Ziegler, 120–121. Helsinki: The Academy of Fine Arts at the University of the Arts, n.d.

Neyrat, Frédéric. "Planetary Antigones: The Environmental Situation and the Wandering Condition." *Qui Parle* 25, nos. 1–2 (2016): 35–64.

Neyrat, Frédéric. "Eccentric Earth." *Diacritics* 45, no. 3 (2017): 4–21.

Neyrat, Frédéric. "Elements for an Ecology of Separation: Beyond Ecological Constructivism." In *General Ecology: The New Ecological Paradigm*, edited by Erich Hörl, translated by James Burton, 101–128. London: Bloomsbury Academic, 2017.

Neyrat, Frédéric. *The Unconstructable Earth: An Ecology of Separation.* Translated by Drew S. Burk. New York: Fordham University Press, 2019.

Nixon, Rob. *Slow Violence and the Environmentalism of the Poor.* Cambridge, MA: Harvard University Press, 2013.

Nuwer, Rachel. "Future Fossils: Plastic Stone." *New York Times*, June 9, 2014.

Oenen, Gijs van. "Interpassive Agency: Engaging Actor-Network-Theory's View on the Agency of Objects." *Theory & Event* 14, no.2 (2011): 1–19.

Oiticica, Hélio. "The Transition of Color from the Painting into Space and the Meaning of Construction." In, *Hélio Oiticica: The Body of Color*, Mari Carmen Ramirez, 222–230. London: Tate Publishing, 2007.

Oiticica, Hélio. "Notes on the *Parangolé*." In *Hélio Oiticica*, 86–89 Paris: Galerie Nationale Jeu de Paume, 1992.

Oliveira, Eduardo Jorge de. "How to Build Cathedrals. Cildo Meireles: A Sensory Geography of Brazil." *Journal of Latin American Cultural Studies* 28, no. 4 (2019): 607–636.

Oppermann, Serpil. "The Scale of the Anthropocene: Material Ecocritical Reflections." *Mosaic: An Interdisciplinary Critical Journal* 51, no. 3 (2018): 1–17.

Orange, Donna M. *Climate Crisis, Psychoanalysis, and Radical Ethics.* New York: Routledge, 2017.

O'Reilly Tonks, Patrick. "Cannibal Routes: Mapping the Atlantic as a Network of Appropriations." PhD diss., University of Michigan, 2013.

Paiva, João Carlos, Carla Morais, and Luciano Moreira. "Specialization, Chemistry, and Poetry: Challenging Chemistry Boundaries." *Journal of Chemical Education* 90, no. 12 (2013): 1577–1579.

Paoletti, Michele Paolini. "Structures as Relations." *Synthese* 195 (2018): 1–20. https://doi .org/10.1007/s11229-018-01918-8.

Parikka, Jussi. *The Anthrobscene*. Minneapolis: University of Minnesota Press, 2014.

Parikka, Jussi. *A Geology of Media*. Minneapolis: University of Minnesota Press, 2015.

Parkes, Alex. "On the Properties of Parkesine, and Its Application to the Arts and Manufactures." *Journal of the Society of Arts* 14, no. 683 (1865): 264–271.

Pelkonen, Eeva-Liisa. "Plastic Imagination." *Forty-Five: A Journal of Outside Research* 2015. http://forty-five.com/media/export/entry_28_lo5.pdf.

Peplow, Mark. "The Plastics Revolution: How Chemists Are Pushing Polymers to New Limits." *Scientific American*. August 21, 2016. https://www.scientificamerican.com /article/the-plastics-revolution-how-chemists-are-pushing-polymers-to-new-limits/.

"Periodic Table of Poetry." Webpage. Accessed September 26, 2021. https://periodictable .group.shef.ac.uk/.

Peters, Kimberly. "Touching the Oceans." *WSQ: Women's Studies Quarterly* 45, nos. 1–2 (2017): 278–281.

Peyton, F. "When Drives Are Dangerous: Drive Theory and Resource Over-Consumption." *Modern Psychoanalysis* 29 (2004): 251–270.

Phillips, Catherine. "Discerning Ocean Plastics: Activist, Scientific, and Artistic Practices." *Environment and Planning A: Economy and Space* 49, no. 5 (May 2017): 1146–1162.

Phillips, Charles. "Bend, Engage, Wait, and Watch: Rethinking Political Agency in a World of Flows." PhD diss., Johns Hopkins University, 2014.

Phillips, John. "Agencement/Assemblage." *Theory, Culture and Society* 23, nos. 2–3 (2006): 108–109.

Pitcher, Wallace Spencer. *The Nature and Origin of Granite*. Dordrecht, NL: Springer Netherlands, 2012.

"Plastic Ocean: The Great Pacific Garbage Patch." Reset.org. Accessed November 22, 2020. https://en.reset.org/knowledge/plastic-ocean-great-pacific-garbage-patch.

"Polymers." Webpage. Accessed November 22, 2020. https://www2.chemistry.msu.edu /faculty/reusch/VirtTxtJml/polymers.htm.

"Polymers and Plastics." AZO Materials. February 14, 2017. https://www.azom.com /article.aspx?ArticleID=13567.

"Polymer Properties." Webpage. Accessed November 22, 2020. http://employees.csbsju .edu/cschaller/Advanced/Polymers/CPxtal.html.

Potgieter, Frederick. "Critique of Relational Aesthetics and a Poststructural Argument for Thingly Representational Art." *Cogent Arts & Humanities* 5, no. 1 (2018). https:// doi.org/10.1080/23311983.2018.1531807.

Povinelli, Elizabeth. *Geontologies: A Requiem to Late Liberalism*. Durham, NC: Duke University Press, 2016.

Povinelli, Elizabeth. "Three Figures of Geontology." In *Anthropocene Feminism*, edited by Richard Grusin, 49–64. Minneapolis: University of Minnesota Press, 2017.

Power, S. "Bound Objects and Blurry Boundaries: Surrealist Display and (Anti)Nationalism." *Journal of Surrealism and the Americas* 2, no. 1 (2008): 95–113.

Powers, Richard. *Gain*. New York: Farrar, Straus & Giroux, 1998.

Probyn, Elspeth. *Eating the Ocean*. Durham, NC: Duke University Press, 2016.

Probyn, Elspeth. "Eating/Space/Media." *Geoforum* 84 (2017): 243–244.

Probyn, Elspeth. "The Ocean Returns: Mapping a Mercurial Anthropocean." *Social Science Information* 57, no. 3 (2018): 386–402.

Proctor, Robert, and Londa L. Schiebinger, eds. *Agnotology: The Making and Unmaking of Ignorance.* Stanford, CA: Stanford University Press, 2008.

"Properties of Polymers." Last updated August 6, 2019. https://chem.libretexts.org/Courses/University_of_British_Columbia/UBC_CHEM_154%3A_Chemistry_for_Engineering/05%3A_Polymers/5.09%3A_Properties_of_Polymers.

Provencher, J. F., J. C. Vermaire, S. Avery-Gomm, B. M. Braune, and M. L. Mallory. "Garbage in Guano? Microplastic Debris Found in Faecal Precursors of Seabirds Known to Ingest Plastics." *Science of the Total Environment* 644 (December 10, 2018): 1477–1484.

Prud'homme, Johanne, and Lyne Légaré. "Semanalysis: Engendering the Formula." In *Signo: Theoretical Semiotics on the Web*, directed by Louis Hébert. Quebec: Rimouski, 2006.

Queyras, Sina. "In Conversation: Sina Queyras & Adam Dickinson." *Lemon Hound*, September 23, 2013. https://lemonhound.com/2013/09/27/in-conversation-sina-queyras-adam-dickinson/.

Raffoul, François, and David Pettigrew. "Translators' Introduction." In Jean-Luc Nancy, *The Creation of the World or Globalization*, translated by François Raffoul and David Pettigrew. Albany: State University of New York Press, 2007.

Railton, Peter. "That Obscure Object, Desire." *Proceedings and Addresses of the American Philosophical Association* 86, no. 2 (November 2012): 22–46.

Ramirez, Mari Carmen, ed. *Hélio Oiticica: The Body of Color.* London: Tate Publishing, 2007.

Rappert, Brian. "Present Absences: Hauntings and Whirlwinds in '-Graphy'." *Social Epistemology: A Journal of Knowledge, Culture and Policy* 28, no. 1 (2014): 41–55.

Rathje, William, and Cullen Murphy. *Rubbish!: The Archaeology of Garbage.* New York: Harper Collins, 1992.

Raghavan, V. "Sahitya." In *An Introduction to Indian Poetics*, edited by V. Raghavan and Nagendra. Bombay: Macmillan, 1970.

Reno, Joshua Ozias. "Toward a New Theory of Waste: From 'Matter out of Place' to Signs of Life." *Theory, Culture and Society* 31, no. 6 (2014): 3–27.

"Rheology." Lexico.com. Accessed September 26, 2021. https://www.lexico.com/en/definition/rheology.

Rhoads, Bruce L., and Colin E. Thorn. "Toward a Philosophy of Geomorphology." In *The Scientific Nature of Geomorphology: Proceedings of the 27th Binghamton Symposium in Geomorphology Held 27–29 September 1996*, edited by Bruce L. Rhoads and Colin E. Thorn, 115–143. New York: John Wiley & Sons, 1996.

Rhodes, Gale. "The Well-Read Biochemist." *Journal of Chemical Education* 73, no. 8 (1996): 732.

Rieser, Max. "Metaphoric Expression in the Plastic Arts." *Journal of Aesthetics and Art Criticism* 17, no. 2 (December 1958): 194–200.

Robbins, Paul, and Sarah A. Moore. "Ecological Anxiety Disorder: Diagnosing the Politics of the Anthropocene." *Cultural Geographies* 20, no. 1 (2013): 3–19.

Roberts, Jody A. "Reflections of an Unrepentant Plastiphobe: Plasticity and the STS Life." *Science as Culture* 19, no. 1 (2010): 101–120.

Rodriguez, Ferdinand. "Plastic: Chemical Compound." *Encyclopedia Britannica*. Last updated November 10, 2020. https://www.britannica.com/science/plastic.

Roelvink, Gerda, and Magdalena Zolkos. "Climate Change as Experience of Affect." *Angelaki* 16, no. 4 (2011): 43–57.

Ronda, Margaret. "Anthropogenic Poetics." *Minnesota Review* 83 (2014): 102–111.

Rosenberg, Harold. *The Anxious Object*. Chicago: University of Chicago Press, 1966.

Ross, James F. *Portraying Analogy*. Cambridge: Cambridge University Press, 2009.

Rosso, Stefano, and Carolyn Springer. "A Correspondence with Umberto Eco: Genova-Bologna-Binghamton-Bloomington August–September 1982, March–April 1983." *boundary 2* 12, no. 1 (1983): 1–13.

Roszak, T. *The Voice of the Earth: An Exploration of Ecopsychology*. MI: Phanes Press, 2001.

Rothko, Mark. *The Artist's Reality: Philosophies of Art*. New Haven, CT: Yale University Press, 1998.

Rothschild, David de. *Plastiki: Across the Pacific on Plastic*. London: Chronicle Books, 2011.

Roy, Sumana. "Untouchability." In *The Mother's Lover and Other Stories*. New Delhi: Bloomsbury, 2020.

Rozwadowski, Helen M. "Technology and Ocean-Scape: Defining the Deep Sea in Mid-Nineteenth Century." *History and Technology, an International Journal* 17, no. 3 (2001): 217–247.

Rozwadowski, Helen M. *Fathoming the Ocean: The Discovery and Exploration of the Deep Sea*. Cambridge, MA: Harvard University Press, 2005.

Rudosky, Christina Helena. "Breton the Collector: A Surrealist Poetics of the Object." PhD diss., University of Colorado, 2015.

Ruy, David. "Review of Eeva-Liisa Pelkonen, 'Plastic Imagination,'" *Forty-Five: A Journal of Outside Research* 2015. http://forty-five.com/media/export/entry_28_lo5.pdf.

Rydz, Evelyn. "Folded Waters." Accessed September 27, 2021. https://evelynrydz.com/section/488627-Folded-Waters.html.

Sallis, John. *Elemental Discourses*. Bloomington: Indiana University Press, 2018.

Salmon, E. "Kincentric Ecology: Indigenous Perceptions of the Human-Nature Relationship." *Ecological Applications* 10 (2000): 1327–1332.

Sampson, Tony D. *Virality*. Minneapolis: University of Minnesota Press.

Scanlan, John. *On Garbage*. London: Reaktion, 2005.

Schaag, Katie. "Plastiglomerates, Microplastics, Nanoplastics: Toward a Dark Ecology of Plastic Performativity." *Performance Research* 25, no. 2 (2020): 14–21.

Schafer, Rieke. "Historicizing Strong Metaphors: A Challenge for Conceptual History." *History of Concepts* 7, no. 2 (2012): 28–51.

Schaller, Chris P. "Polymer Properties." In *Structure and Reactivity in Organic, Biological and Inorganic Chemistry: An Online Chemistry Textbook*, undated. http://employees.csbsju.edu/cschaller/Advanced/Polymers/PolyProperties.html/.

Scheffler, Israel. *Philosophy and Education*. Boston: Allyn and Bacon, 1966.

Schummer, Joachim. "Aesthetics of Chemical Products: Materials, Molecules, and Molecular Models." *Hyle–International Journal for Philosophy of Chemistry* 9, no.1 (2003): 73–104.

"Science Matters: The Case of Plastics." Science History. Accessed November 22, 2020. https://www.sciencehistory.org/science-of-plastics.

Scott, Charles. *Living with Indifference*. Bloomington: Indiana University Press, 2007.

Scovino, Felipe. "Tactics, Positions and Inventions: Devices for a Circuit of Irony in Contemporary Brazilian Art." *Third Text* 21, no. 4 (2007): 431–440.

Scully, Matthew. "Plasticity at the Violet Hour: Tiresias, *The Waste Land*, and Poetic Form." *Journal of Modern Literature* 41 no. 3 (2018): 166–182.

Seamon, David, and Arthur Zajonc, eds. *Goethe's Way of Science: A Phenomenology of Nature*. New York: State University of New York Press, 1998.

Searles, H. *The Nonhuman Environment in Normal Development and Schizophrenia*. New York: International Universities Press, 1960.

Searles, Harold F. "Unconscious Processes in Relation to the Environmental Crisis." *Psychoanalytic Review* 59 (1972): 361–374. http://collectivetask.com/wp-content/uploads/2017/12/5_04_rosenfield.pdf.

Semper, Gottfried. "The Four Elements of Architecture." In *Architectural Theory: Volume 1—An Anthology from Vitruvius to 1870*, edited by Harry Francis Mallgrave, 536–539. Malden, MA: Blackwell, 2006.

Serres, Michel. *The Birth of Physics*. Translated by Jack Hawkes. Manchester: Clinamen Press, 2000.

Shapiro, Nicholas. "Attuning to the Chemosphere: Domestic Formaldehyde, Bodily Reasoning and the Chemical Sublime." *Cultural Anthropology* 30, no. 3 (2015): 368–393.

Shapshay, Sandra. "Contemporary Environmental Aesthetics and the Neglect of the Sublime." *British Journal of Aesthetics* 53, no. 2 (April 2013): 181–198.

Shaw, I., and K. Meehan. "Force-Full: Power, Politics and Object-Oriented Philosophy." *Area* 45, no. 2 (June 2013): 216–222.

Shellhorse, Adam Joseph. "Subversions of the Sensible: The Poetics of Antropofagia in Brazilian Concrete Poetry." *Revista Hispánica Moderna* 68, no. 2 (2015): 165–190.

Sheng, James J. *Modern Chemical-Enhanced Oil Recovery: Theory and Practice*. Oxford: Gulf Professional Publishing, 2011.

Shimony, Abner. *Search for a Naturalistic World View*, Vol. 1. Cambridge: Cambridge University Press, 1993.

Shotwell, A. "Water and Ocean." *Cultural Studies Review* 23, no. 2 (2017): 184–188.

Silva, Renato Rodrigues da. "Hélio Oiticica's *Parangolé* or the Art of Transgression." *Third Text* 19, no. 3 (2005): 213–231.

Silverman, H. J. "Malabou, Plasticity, and the Sculpturing of the Self." *Concentric: Literacy & Cultural Studies* 36 (2010): 89–102.

Skrebowski, Luke. "Revolution in the Aesthetic Revolution: Hélio Oiticica and the Concept of Creleisure." *Third Text* 26, no. 1 (2012): 65–78.

Skrzypek, Jeremy W. "From Potency to Act: Hyloenergeism." *Synthese* (2019). https://doi.org/10.1007/s11229-019-02089-w.

Sloterdijk, Peter. "The Anthropocene: A Process-State at the Edge of Geohistory?" In *Art in the Anthropocene: Encounters among Aesthetics, Politics, Environments and Epistemologies*, edited by Heather Davis and Etienne Turpin, 327–340. London: Open Humanities Press, 2015.

Solnit, Rebecca. "Water." In *A Companion to American Environmental History*, edited by Douglas Cazaux Sackman, 92–96. Chichester, West Sussex: Blackwell, 2010.

Souriau, Etienne. "Time in the Plastic Arts." *Journal of Aesthetics and Art Criticism* 7, no. 4 (1949): 294–307.

Spurlock, Katherine. "Rival Authorities: Sigmund Freud, T. S. Eliot and the Interpretation of Culture." PhD diss., University of Virginia, 1997.

Starosielski, N. *The Undersea Network*. Durham, NC: Duke University Press, 2015.

Stauffer, Eric, Julia A. Dolan, and Reta Newman, *Fire Debris Analysis*. Burlington, MA: Academic Press, 2008.

Stayer, Jayme. "The Dialogics of Modernism: A Bakhtinian Approach to T. S. Eliot's *The Waste Land* and Igor Stravinsky's *Oedipus Rex*." PhD diss., University of Toledo, 1995.

Steinberg, Philip E. "Of Other Seas: Metaphors and Materialities in Maritime Regions." *Atlantic Studies* 10, no. 2 (2013): 156–169.

Stewart, Kathleen. "Afterword: Worlding Refrains." In *The Affect Theory Reader*, edited by M. Gregg and G. Seigworth, 339–354. Durham, NC: Duke University Press, 2010.

Stiegler, Bernard. "General Ecology, Economy, and Organology." In *General Ecology: The New Ecological Paradigm*, edited by Erich Hörl, translated by Daniel Ross, 129–150. London: Bloomsbury Academic, 2017.

Stolorow, R. "Climate Change, Narcissism, Denial, Apocalypse." *Psychology Today*, October 6, 2012. https://www.psychologytoday.com/ blog/feeling-relating-existing/201210 /climate-change-narcissism-denial-apocalypse.

Stolorow, R. "Death, Afterlife and Doomsday Scenario." *Psychology Today*, December 10, 2013. https://www.psychologytoday.com/us/blog/feeling-relating-existing/201312 /death-afterlife-and-doomsday-scenario.

Strang, Veronica. "Introduction: Toward a Hydrological Turn?" In *Thinking with Water*, edited by Cecilia Chen, Janine MacLeod, and Astrida Neimanis. Montreal: McGill-Queen's University Press, 2013.

Sugars, Cynthia. "The Evolutionary Sublime: Deep Time and the Historical Novel in Joan Thomas's Curiosity." *Mosaic: An Interdisciplinary Critical Journal* 51, no. 3 (2013): 199–221.

Sullivan, Heather I. "Goethe, the Romantics and Early Geology." *European Romantic Review* 10, nos.1–4 (1999): 341–370.

Taffel, Sy. "Technofossils of the Anthropocene: Media, Geology, and Plastics." *Cultural Politics* 12, no. 3 (2016): 355–375.

Tagore, Rabindranath. "Sāhitya Shristi." *Rabindra Rachanavali*, 341–350 Calcutta: Government of West Bengal, 1989.

Tagore, Rabindranath. "Visva Sahitya." In *Rabindranath Tagore in the 21st Century: Theoretical Renewals*, edited by Debashish Banerji, translated by Rijula Das and Makarand R. Paranjape, 277–288 (New Delhi: Springer, 2015).

Taylor, Matthew. "The Nature of Fear: Edgar Allan Poe and Posthuman Ecology." *American Literature* 84, no. 2 (2012): 353–379.

"Tess Felix—Curious Remnants: An Ocean in Crisis." Peninsula Museum of Art. Accessed November 23, 2020. https://www.peninsulamuseum.org/exhibits/curious -remnants-ocean-crisis#.XwvtK5gzbIU.

Teuten, E. L., Jovita M. Saquing, Detlef R. U. Knappe, Morton A. Barlaz, Susanne Jonsson, Annika Björn, Steven J. Rowland, Richard C. Thompson, Tamera S. Galloway,

Rei Yamashita, Daisuke Ochi, Yutaka Watanuki, Charles Moore, Pham Hung Viet, Touch Seang Tana, Maricar Prudente, Ruchaya Boonyatumanond, Mohamad P. Zakaria, Kongsap Akkhavong, Yuko Ogata, Hisashi Hirai, Satoru Iwasa, Kaoruko Mizukawa, Yuki Hagino, Ayako Imamura, Majua Saha, and Hideshige Takada. "Transport and Release of Chemicals from Plastics to the Environment and to Wildlife." *Philosophical Transactions* 364, no. 1526 (July 27, 2009). https://www.ncbi.nlm.nih.gov/pmc/articles/PMC2873017/.

Tezuka, Miwako. "Jikken Kōbō (Experimental Workshop): Avant-Garde Experiments in Japanese Art of the 1950s." PhD diss., Columbia University, 2005.

Tezuka, Miwako. "Experimentation and Tradition: The Avant-Garde Play *Pierrot Lunaire* by Jikken Kōbō and Takechi Tetsuji." *Art Journal* 70, no. 3 (2011): 64–85.

Thomas, Julia Adeney. "History and Biology in the Anthropocene: Problems of Scale, Problems of Value." *American Historical Review* 119, no. 5 (2014): 1587–1607.

Thompson, Richard C. "Plastics, Environment and Health." In *Accumulation: The Material Politics of Plastic*, edited by J. Gabrys, G. Hawkins, and M. Michael, 150–168. London: Routledge, 2013.

Thompson, Richard C., Shanna H. Swan, Charles J. Moore, and Frederick S. vom Saal, "Our Plastic Age." *Philosophical Transactions of the Royal Society B: Biological Sciences* 364 (2009): 1973–76.

Toadvine, Ted. "The Elemental Past." *Research in Phenomenology* 44, no. 2 (2014): 262–279.

Tomkins, Silvan S. *Affect Imagery Consciousness: The Complete Edition*. New York: Springer, 2008.

Uhlmann, D. R., and A. G. Kolbeck. "The Microstructure of Polymeric Materials." *Scientific American* 233, no. 6 (December 1975): 96–107.

Ulmer, Jasmine B. "Plasticity: A New Materialist Approach to Policy and Methodology." *Educational Philosophy and Theory* 47, no. 10 (2015): 1096–1109.

Valentine, Ben. "Plastiglomerates: The Anthropocene's New Stone." Hyperallergic, November 25, 2015. https://hyperallergic.com/249396/plastiglomerate-the-anthropocenes-new-stone/.

Walker, Anthony. "Plastics: The Building Blocks of the Twentieth Century." *Construction History* 10 (1994): 67–88.

Wallace, Molly. *Risk Criticism: Precautionary Reading in an Age of Environmental Uncertainty*. Ann Arbor: University of Michigan Press, 2016.

Wallance, Donald A. "Design in Plastics." *Everyday Art Quarterly*, no. 6 (1947–1948): 3–15.

Weber, Andreas. *Enlivenment: Towards a Poetics of the Anthropocene*. Cambridge, MA: MIT Press, 2019.

Weininger, Stephen J. "Butlerov's Vision: The Timeless, the Transient, and the Representation of Chemical Structure." In *Of Minds and Molecules: New Philosophical Perspectives on Chemistry*, edited by Nalini Bhushan and Stuart Rosenfeld, 143–161. New York: Oxford University Press, 2000.

Weintrobe, Sally, ed. *Engaging with Climate Change: Psychoanalytic and Interdisciplinary Perspectives*. New York: Routledge, 2013.

Wernick, Iddo K., Robert Herman, Shekhar Govind, Jesse Ausubel, and H. Daedalus. "Materialization and Dematerialzation: Measures and Trends." *Boston* 125, no. 3 (1996): 171–198.

Whitehead, Mark. *Environmental Transformations: A Geography of the Anthropocene.* London: Routledge, 2014.

Williams, Tyler. "Plasticity, in Retrospect: Changing the Future of the Humanities." *Diacritics* 41, no. 1 (2013): 6–25.

Williston, Byron. "The Sublime Anthropocene." *Environmental Philosophy* 13, no. 2 (2016): 155–174.

Wilson, Jonathan. "Primo Levi's Hybrid Texts." *Judaism: A Quarterly Journal of Jewish Life and Thought* 48, no. 1 (1999): 67–72.

Wujcik, Stacey A. "An Anthropophagic Legacy: Oswald de Andrade's Manifesto Antropofago in Brazilian Anti-Art and the Works of Cildo Meireles." PhD diss., Temple University, 2011.

Yaeger, Patricia. "The Death of Nature and the Apotheosis of Trash; or, Rubbish Ecology." *PMLA* 123, no. 2 (2008): 321–339.

Yaeger, Patricia. "Sea Trash, Dark Pools, and the Tragedy of the Commons." *PMLA* 125, no. 3 (2010): 523–545.

Yarsley, E., and E. G. Couzens. *Plastics.* Hammondsworth, UK: Penguin, 1941.

Yarsley, V. E., and E. G. Couzens. "The Expanding Age of Plastics." *Science Digest* 10 (1941): 57–59.

Yoldas, Pinar. "Plastisphere Lexicon." Ecosystem of Excess, 2014. Accessed September 27, 2021. https://cargocollective.com/yoldas/WORK/Ecosystem-of-Excess-2014.

Yoldas, Pinar. "Speculative Biologies: New Directions in Art in the Age of the Anthropocene." PhD diss., Duke University, 2016.

Yusoff, Kathryn. "Geologic Life: Prehistory, Climate, Futures in the Anthropocene." *Environment & Planning D: Society & Space* 31, no. 5 (2013): 779–795.

Yusoff, Kathryn. "Geologic Subjects: Nonhuman Origins, Geomorphic Aesthetics and the Art of Becoming Inhuman." *Cultural Geographies* 22, no. 3 (2015): 383–407.

Zalasiewicz, Jan. *The Earth after Us: What Legacy Will Humans Leave in the Rocks?* New York: Oxford University Press, 2008.

Zalasiewicz, Jan. *The Planet in a Pebble: A Journey into Earth's Deep History.* Oxford: Oxford University Press, 2010.

Zielinski, Siegfried. *Deep Time of the Media: Toward an Archaeology of Hearing and Seeing by Technical Means.* Translated by Gloria Custance. Cambridge, MA: MIT Press, 2005.

Zilberman, David B. *Analogy in Indian and Western Philosophical Thought,* edited by Helena Gourko, and Robert S. Cohen. Heidelberg: Springer, 2006.

Zimmer, Carl. "Hypersea Invasion." *Discover Chicago* 16, no. 10 (October 1995): 76–87.

Zivin, Erin Graff. *Anarchaeologies: Reading as Misreading.* New York: Fordham University Press, 2020.

Žižek, Slavoj. *The Ticklish Subject: The Absent Centre of Political Ontology* [1999]. London: Verso, 2008.

Index